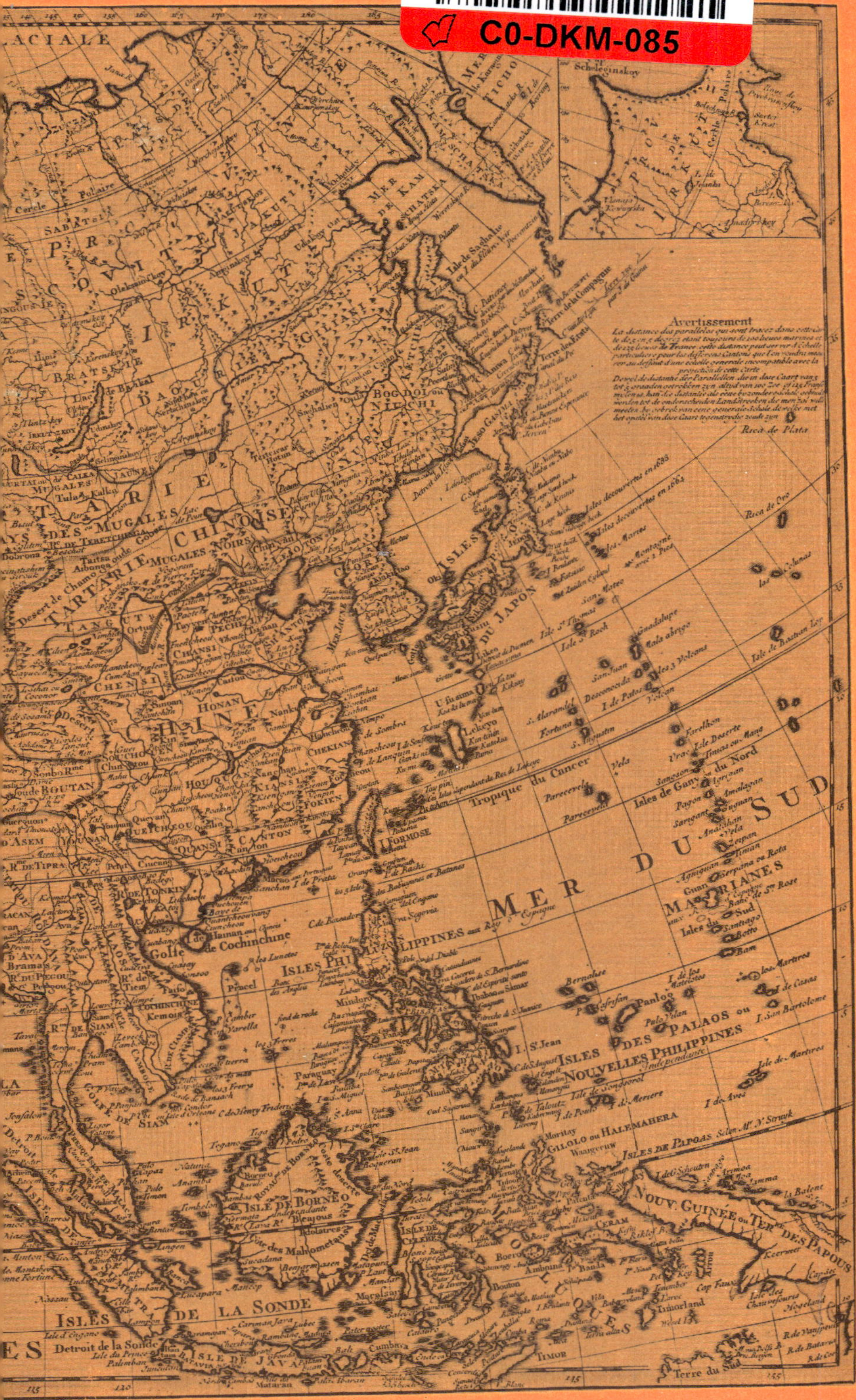

ACIALE

TARTARIE CHINOISE

TARTARIE DES MUGALES

TANGUT

CHINE

HONAN

CHENSI

CHEKIA

HOUQUANG

KIANSI

FOKIEN

QUANSI

CANTON

QUEICHOU

YUNAN

TONKIN

COCHINCHINE

SIAM

ISLES PHILIPPINES

FORMOSE

MER DU JAPON

MER DU SUD

Tropique du Cancer

MARIANES

ISLES DES PALAOS ou NOUVELLES PHILIPPINES

ISLES DE PAPOS

GILOLO ou HALEMAHERA

NOUV. GUINEE ou TERRE DES PAPOUS

ISLE DE BORNEO

ISLES DE LA SONDE

Detroit de la Sonde

ISLE DE JAVA

TIMOR

Terre du Sud

Avertissement

Recit de Plata

John Kenneth Galbraith
Introduces India

*A female custodian of the 13th century Sun
Temple (Black Pagoda) at Konarak.*

John Kenneth Galbraith

introduces

INDIA

Edited by Frank Moraes and Edward Howe

ANDRE DEUTSCH

First published 1974 by
André Deutsch Limited
105 Great Russell Street London WC1

Copyright © 1974 by André Deutsch Limited
'Introducing India' copyright © 1974 by John Kenneth Galbraith
All rights reserved

Printed in Hong Kong by Leefung-Asco Printers Ltd.

ISBN 0 233 96218 2

Grateful acknowledgment is made to Messrs Weinreb & Douwma Ltd, map dealers
of London, for the loan of the map reproduced as endpapers to this book. We also
owe thanks to the Government of India Tourist Office in London for their kindness
and help in loaning and identifying some of the illustrations. The photograph of
the tiger is by Peter Jackson; the photograph of the gateway to the Great Stupa at
Sanchi is by Bury Peerless; the photograph of the Rajasthani bridegroom on a camel
is by Richard Holkar; and the following photographs are by André Deutsch: the
Taj Mahal, on the back of the jacket; details from the stone friezes at Mahabalipuram; the shore temple at Mahabalipuram; the Meenakshi Temple at Madurai
and the canopy in the Meenakshi Temple complex, Madurai.

Contents

APPENDIX

Illustrations

Introducing India

John Kenneth Galbraith

Ever since I served in India as an ambassador I have function-
ed as a minor oracle and guide for an appreciable fraction of
all the people contemplating a journey to that marvellous and
quite mysterious land. I decided some years ago that I should
codify my advice. This is it. The rather simple point made here
is what most aided my own understanding. I doubt that it is
terribly original; most knowledge, in this day, involves not
invention but sorting out from what other people believe the
things that you should believe yourself.

The first question one is asked by the intending visitor in-
vokes the central point. The question is: 'What should I see?'
This inquiry is usually made in connection with an intended
visit of two or three weeks: sometimes it is made in anticipation of a stay of two or three days although then, usually, with
the half-apologetic addition: 'I realise, of course, that I can't
see *everything* in that time.'

Once, no doubt, Americans visiting Europe or Englishmen
departing for the continent asked the same question. But that
must have been some time back. It is not a question that one
can ask about a continent and it is the beginning of wisdom
about India to realise that it is not a country but a continent.
Within it, between greatly blurred boundaries, are many nations.

Geographically the point is beyond dispute. The contrast
between the Himalayas and the plains of the Ganges and the
Punjab is a good deal greater than that between the Alps and
(say) the south of Spain, and much more abrupt. There is high
country in south central India that is like England and not far
from a town called Wellington one can go riding on the
downs. Nearby is a terrifying hill with numbered hairpin
turns; on the way down are coffee plantations and at the bot-

1

tom rubber trees, rice paddies and palms. On the west of India is a great desert from which the sand blows for hundreds of miles in the spring to darken and, oddly enough, to cool Delhi. Bengal, which is to the east, is only barely above water and in Assam just north, where it rains every hour or so, several hundred inches fall in the course of a year.

But differences in geography are easy enough to grasp. A country can accommodate an unlimited number and still qualify as a country. It is the differences in language, social and economic behaviour, art and architecture and religion that make India many nations. Since long before history became reliable, invaders have been sweeping into the subcontinent from the north and west. Some overran nearly the whole country (the far south was usually spared) and remained; some made brief incursions and departed. The imperial power of those that remained was rarely durable; the central authority soon weakened and imperial satrapies and feudatories became independent states.

India's rulers have rarely been permissive at least about the things they thought important. (Measured by results they were right. The whole business of religious conversion went into a great decline when those in charge ceased to use death and eternal damnation as a threat and education, medicine and food as a bribe.) In consequence, the invaders implanted their preferences as to language and religion in the parts of the country that they governed. And their courts reflected their preferences in art and architecture and their feudatories and satraps implanted *their* personality although in more imitative fashion.

The last of the inland invaders, the Mughals, came in from Central Asia and deposed the previous intruders in 1526. They were among the most inspired builders and patrons of the arts of all time. The contrast between their palaces, tombs and mosques in central and northern India and the temples of the older Hindu tradition in the south is especially sharp.

On the swords of the invaders came Islam. And adding to the durable and still poignant diversity which this brought to the subcontinent were also the effects of the great indigenous religious movements – those of the Buddhists, Jains, Sikhs and Christians. Like the military invaders, religion moved

2

over India with varying effect and varying capacity to survive. To the diverse tradition of courts and rulers was added that of faith. The relevant history is, I think, admirably summarised in *India: Now and Through Time* by Catherine Galbraith and Rama Mehta (New York: Dodd Mead, 1971). Last of all, from over the sea, came the Europeans and most notably the British. They were the first invaders in historical times to establish hegemony over the whole continent; it is mostly owing to the British that we think of India as one country rather than a federation of many. Although the Europeans left their mark on the design and architecture of the big cities – New Delhi, Bombay, Calcutta and Madras – this westernisation would doubtless have been accomplished anyway, in time, by Intercontinental, Caltex and the automobile. The British cultural impact, which was political and moral rather than artistic, was also horizontal rather than vertical. It captured not a region but a stratum. To this important point I will return.

Thus the many nations of India. Some are reasonably close in language, art, architecture and tradition – as close, perhaps, as France and Italy. Such might be a fair description of the relation between the great northern state of Uttar Pradesh, centering on Lucknow and embracing the old Muslim kingdom of Oudh, and the neighbouring states of Bihar and Bengal. But Madras, Mysore and Kerala in the south are culturally much more distant from Uttar Pradesh than Germany from Spain. Germans and Spaniards use a roughly similar alphabet; the language of the south of India is as mysterious to a northerner as Arabic to an Austrian. The difference between the Alhambra and the cathedral in Cologne is not greater than that between the Red Fort in Delhi and the great temples of the south.

No less important is the difference in social temper and economic organisation as between the Indian nations. Most Indians make their living – or survive – by scratching it from the soil. The Indian village has the dignity and stability of an old culture. But life for most of its inhabitants is meagre and brief. But not everywhere. In the Punjab in the north-west farmers ride tractors, use fertilizers and grow abundant crops of the new high-yielding varieties of wheat. On one thing

Hindu, Sikh, Christian and Voodoo gods agree. It is that in practice as distinct from principle those who have the most should get the most. In consequence of this rule and their own intelligent efforts the Punjabis now have a standard of living, and in any case a mode of life, that is probably superior to that of the small farmer in the Appalachian plateau in the United States. They supply to the rest of India the wheat that once came from Kansas, the Dakotas and Washington. Bengalis and Madrasis are much poorer and another comparison with Europe is apt. For, as in the case of Greeks and Italians contemplating the Swiss or the Swedes, they concede the northerners their vastly superior industry and modest but still greatly superior wealth. The Bengalis and Madrasis insist only on their own more eclectic culture and wisdom and believe that, in consequence, they should write and keep the books and, in considerable measure, run the country.

The meaning of this is that one does not visit India or know India. One visits a part of India – what elsewhere would be one of the nation-states. One should be content to know something of that part; it is better to know something of something than nothing of much. No one is apologetic about trying to understand the Dutch while being happily ignorant of the Irish. Few living Americans or Europeans have travelled over as much of India as I have. It is a great help and a matter of considerable personal economy to move about at public expense in a government plane. For a long while I found myself delighted by my continual sense of discovery. Then I became depressed at how little I still seemed to know. Then I discovered that I was ambassador not to one country but to many. My morale improved. So it will be with the visitor who visits India and comes away feeling that his voyage of discovery never quite began.

I've been speaking of the regional diversity of India. There is also diversity on the vertical plane – a diversity that divides Indians into great, mostly non-communicating layers. To understand this is also most important. I am not speaking of the caste system, the great Indian *cliché*. Indians respond to western curiosity about caste much as do self-righteous Americans when asked severe questions about racial discrimination. Alas, yes, it existed and the attitudes still survive. But progress is being made. I've always thought Indians much too apologetic about caste. The most remarkable feature of Indian cultural, social and economic life has been its genius for survival. Before there were schools or universities, a tradition that required that children be trained or made learned in the occupational knowledge of their parents was not a bad substitute. This the caste system accomplished. But the survival of caste attitudes is not my interest here. By vertical diversity I refer to the line that divides village India, which is most of India, from modern urban, bureaucratic and intellectual India.

Most Indians live in the villages and directly or indirectly draw their living from the land. They pay taxes. Some of the sons go off to the cities or to the army. Almost all listen to music on transistor radios. In proportion to the electoral roll the villagers vote in numbers that make western electorates seem indifferent. Yet the Indian village is not much governed either by New Delhi or the capital of its state. Nor is it governed in any actual sense by itself. It lives by an accepted code of behaviour, not by formal law. This is better. The people of India learned centuries ago to hope for the least possible attention from higher levels of government, for until well into British times all governments, with rare and temporary exceptions, were instruments of spoliation. They sent armies to pillage and tax collectors to extort. The self-sufficiency of the village was its strength; to be left alone was its greatest good fortune. The lesson of centuries is not soon unlearned.

The remainder of India – the upper horizontal slice – is, by comparison, very small. This is the India of the civil service, parliament and the great business houses of Birla and Tata; the India of the Five Year Plans and the steel plants and disputes over the relative rôles of the public and private sector; the India of newspapers, airplanes, armed forces, foreign policy, concern over the United Nations and about Kashmir; of the Chief Ministers of the states and the municipal corporations of Bombay, Calcutta and Madras. In this world, quite literally, everyone of major importance knows every other person by sight and name. One rarely needs to introduce one Indian of this world to another. It is also complete and

5

self-contained – so much so that most foreigners and a great many Indians come to think of it as all of India. For understanding India that is the greatest mistake of all.

For eventually one comes to realise that much can happen in this world while nothing changes in India. Atomic power plants are built; the bomb is debated; a new Five Year Plan is discussed; a treaty is concluded with the Soviet Union; across the desks of the civil servants move great piles of files, the paper grey and dog-eared. The fate of India is not involved. It would be a mistake to dismiss modern India; it has vitality, an abounding and sensitive pride, an undoubted undercurrent of self-righteousness, and a marked capacity for self-deprecation and humour. But its wheels turn by themselves.

There are, to be sure, points where the two Indias come together. The fertilizers and hybrids which are making the green revolution are the result of such a conjunction. So was the virtual elimination of malaria, an accomplishment which would cause a good many people to enter a minority vote on behalf of DDT. There are also other places where the future of India depends on effecting a conjunction. This is notably so in the case of birth control. This comes from the modern sector – the upper slice. So far it has not crossed the chasm. And it is in the villages that the babies are born. No fact concerning India is so important – so all-embracing – as the resulting relentless expansion in population. It is the ultimate constraint on every chance for escape from the rending poverty that is on every hand.

The regional diversity of India is obtrusive. Even the shortest stay will sufficiently persuade the visitor as to how much more there is to see and to learn. The vertical diversity is seductive: it causes people to imagine that, having experienced the small world of modern India, they have experienced all of India. As I have said, this is an error common to both foreigners and many Indians. I had been long in Delhi before I realised how urgent could be the discussion of economic planning, village development, schemes for health and educational betterment, development of village crafts and, of course, family planning and how slight would be the consequence. The discussion was an art form of the modern slice – or, at best, an expression of its conscience and concern. The reality was the absence of any levers for moving the great village mass – the absence, on occasion, even of means of communication with the India of the millions.

Thus the several and diverse nations of India. It is no small task, this effort at comprehension on which the reader is now embarked. I trust that this framework will slightly aid the task – and slightly ease the pain when the reader discovers how much he has studied and how much he has still to learn. Whatever happens, if my own experience is any guide, it will be an enchanting adventure. All good luck.

Travel in India

Edward Howe

A 'long walk' from Aleppo in Syria to Ajmer in Rajasthan, a distance of some 3,300 miles, made Thomas Coryat the first English traveller or tourist to India. He undertook the trip, for no other reason than curiosity, in 1614–15. A few years earlier this amazing traveller published a book called *Coryat Crudities* which described his European journeys. The pity is that he did not live to write of his journey to India or of his travels on the subcontinent. His ten-month-long walk cost him fifty shillings.

Curiosity is certainly the best incentive for travel, and especially for a trip to India. Whether you class yourself as a tourist or a traveller, anyone visiting India should approach his tour by borrowing a little from Indian philosophy, although perhaps patience is a better word. India is a special place, its people have a special conception of life that is totally different from anything anywhere else in the world. It is this difference that compels people to visit this country, a curiosity to try and understand what is this subcontinent, to explore a country that is two-thirds the size of Europe, excluding Russia. The scenery of India is as splendid as it is varied, from the highest mountains in the world, the Himalayas, to the Ganges plain, extending from the Deccan plateau to the lower mountains of the south. In the north east is the wettest region of the world with its tea plantations; in the west are the marshes of Kutch, the haunt of water birds, wild asses and smugglers. Along India's coasts are some of the finest beaches in the world, vast expanses of sand trod only by the fishermen passing from their villages to their boats. Only in recent years are some of these being developed, particularly in Goa, Kerala and Orissa. For those interested in wild life there are game preserves in most regions hiding the elusive and vanish-

8

A general view of the 13th century Sun Temple (Black Pagoda) at Konarak Orissa.

ing tiger, lions – the only ones left in India are carefully preserved in the Gir Forest – bear, elephant and rhinoceros. Flowering trees in their seasons transform towns and roads into places of astonishing beauty. Colourful birds flit everywhere.

Naturally, in a subcontinent like India, many extremes of climate are found, but generally it can be said India's climate is hot, especially so in summer. The higher you climb, the cooler it becomes and these cooler regions are snow-covered in winter. The tourist can choose the climate he wants and does not need to bear the oppressive dry heat of summer in New Delhi or the humid heat of Bombay. Obviously a country embracing mountains and plains, a long coastline, rivers and lakes, and with such a varied climate has produced varying types of people who speak widely different languages, profess a number of religions, and live vastly different lives. A tourist can choose the type of people he wants to meet, as he can the weather.

Today's tourists to India on the whole choose the easiest and the quickest route, by air, which from the furthest part of the world takes less than a day, but costing considerably more than that fifty shillings Thomas Coryat spent just over 350 years ago. The advantage air travel has is that a ticket can include most vantage points for travel in India at no extra, or little extra cost.

For those who visit the subcontinent by sea, Bombay is the main port of entry, but other ports include Calcutta and Madras. Nowadays more travellers, particularly the young, prefer to take the land route. This can take as long as Thomas Coryat took, but for those in a hurry coaches, buses and cars can complete the journey in three weeks from anywhere in Europe. Roads are being improved along the Asian Highway so that a motorist really in a hurry, and with a co-driver, can accomplish the journey in a week – although they would be very tired at the end of such a headlong rush over 7,000 miles. A more leisurely drive is recommended to enable visits to sites of ancient civilisations en route, many with links with similar civilisations as on the subcontinent itself.

As I have stressed, India is a special country; its own special problem is that of over-population from which all

9

op: *Detail of the three-tiered roof of the Audience Hall of the Sun Temple Black Pagoda) at Konarak.* Bottom: *An 18th century Jain Temple in Ahmeda-ad.*

other problems spring. From a tourist point of view none of these problems need be felt although they might be witnessed. He can fly from one air-conditioned hotel to another a thousand miles distant seeing but little of the real India. Modern jets now fly so high the land is lost to view. There is one exception though. Anyone flying from New Delhi to Calcutta in the early morning should make sure he books a seat on the left (port) side of the plane when he can witness one of the most magnificent sights in the world, a panorama of the Himalayas, even though they are about a hundred miles away. Cloud invariably hides this view later in the day.

One of India's problems is that in general travel facilities do not yet approach European or American standards, but the curious traveller venturing off the beaten track, knowing he cannot expect these. can still be very comfortable and by treading a less expensive path he will reap the reward of seeing more of the country and its people. It does, however, need more attention to detail in planning.

I remember an American woman tourist late at night at Calcutta airport crying with deep feeling: 'Oh, I want to go home.' She belonged to that brand of tourist who fly round the world at great speed never really stopping long enough in one place. They miss a lot. India is a country which should be visited for its own sake. Some of the chapters in this book provide pointers on what to see in India. The slip-through tourist is easily advised for he has time only for a few of the highlights. One rather unkind guide described them as 'the walking dead'. After slipping from one country to another they often, like my American woman, want only to go home.

New Delhi is a good stop-over place for all tourists coming to India, providing a good jumping-off centre to take in Agra with the Taj Mahal, and a quick drive from there to see the fascinating deserted city of Fatehpur Sikri. It is a short flight or drive to Jaipur to see the Pink City and the interesting Amber Palace; or to Khajuraho to see the magnificent temples with their fabulous, often erotic carvings. From Delhi too you can fly to Kashmir to taste the pleasure of staying on a houseboat on one of the lakes; or play golf on the highest golf course in the world. One day this region might become the venue for the World Olympic Winter Games.

Should the traveller choose Bombay as his port of arrival, he has a different India to see. Bombay is a bustling port and commercial centre. While it possesses the best hotel in India, the Taj Mahal Hotel, its other buildings have no striking quality except being Victorian mock Gothic. But a number of interesting excursions can be made from here, the nearest being a trip in a launch across the harbour to see the Brahmanical caves on Elephanta Island. Further afield are both Ellora, the home of Buddhist, Hindu and Jain caves, and Ajanta with its monasteries and caves possessing the famous wall paintings. It is worth a stay here, possible at nearby Aurangabad (where there is a hotel) to also see Daulatabad, having first read a little of the dramatic march made by one of Delhi's eccentric sultans, Mohammed Tughluq, who transferred his capital there in 1338. Just outside New Delhi is his former capital, Tughluqabad, a sizeable ruin also well worth a visit.

Also from Bombay an air trip to Goa, or a day's sail by steamer, is interesting especially for those who want to see some of the enormous cathedrals, basilicas and palaces built by the Portuguese in the sixteenth century when they established their colony. While in Bombay you could also take a 119-mile-long train journey on the Deccan Queen up to Poona through the Western Ghats, gaunt rocky hills where some Marathi forts still crown the summits.

A visit to Gujarat is easily made from Bombay, either by air or train, and well worthwhile, for there are varied attractions in this seldom visited state. At Palitana and Junagadh, perched high on the hills, are fascinating citadels of fabulous Jain temples. A tour of the Jain temples, sometimes staying the night in peaceful Jain resthouses at low cost often in the company of pilgrims makes an unusual experience. Gujarat's capital, Ahmedabad, has some of the most beautiful mosques in India, while the art museums in Baroda are famous.

For those whose gateway to India is Madras, yet another India awaits them. In this southern region live the Tamils, a hard-working, disciplined people steeped in Hindu tradition. Here the best dancers and musicians of India receive an exhaustive training in the two arts at the Kalakshetra. There are many Hindu temples to see, and Fort St George, built in 1640

11

by the British, should also be visited.

From Madras it is important to visit Mahabalipuram, 53 miles south along the coast, where there is a collection of rock-hewn monuments including rock carvings of elephants, and to go to the famous temples of Tirupati, Madurai, Tiruchira-palli and Rameswaram will provide a vivid insight into Hin-duism. Madras is also a starting-off point for one of the most interesting corners of India, the state of Kerala (you can also reach it easily from Bombay). This narrow strip of land in the south-west, one of the most beautiful and fertile regions of India, is recognized as having the most highly educated popu-lation in the country. Kerala must be the only place in the world that has voted in, and out, a Communist government. It is a countryside of lakes, creeks and canals, a land of coco-nuts and bananas, of attractive thatched and timbered houses, and, curiously, a great number of churches, for this is also the home of the Syrian Christians. There are mountains reaching to over 8,000 feet, virgin jungle where rare orchids flourish, and is the home of elephants and other wild life.

Another two centres reached from Madras are Hyderabad and Bangalore. In the former there are interesting Muslim monuments and it was once a happy hunting ground for anti-que hunters. Mysore, of which Bangalore is the capital, is a state whose history goes back to the days of Asoka, and later became the cradle of dynasties who fostered the arts of sculp-ture and architecture. It is in Mysore that you find some of the best shrines and temples, such as those at Belur, Somnathpur and Halebid. More prosaically, Mysore houses the world's deepest gold mine, at Kolar. Most of Mysore State is at an elevation of over 2,000 feet and is a land of primeval forest where elephants roam and provide the spectacle Kheda (the rounding up of elephants). Pageantry is provided each year in Mysore City at the time of the Dussehra festivities which last for ten days in September/October.

Now we come to the last centre in India for tourists, Cal-cutta, not one of my favourite cities on account of its size and turbulent population. But it has its charms for some, and it is the best centre from which to travel to the temple city of Bhubaneswar. Also Puri, a place of Hindu pilgrimage con-taining the shrine of Jagannath and the scene of the annual

Car Festival when the image of Jagannath is carried on a 'car' that is forty-five feet high and motivated by sixteen wheels each seven feet in diameter. Its fine sandy beach also makes it a popular holiday resort. About 53 miles from Puri is Konarak where the Black Pagoda, the thirteenth-century sun temple, can be seen with its carvings and statuary, the latter claimed to be the best in India.

From Calcutta an overnight rail journey, completed by a 50-mile ride on a light mountain railway, brings you to Darjeeling, and a short pre-dawn walk takes you to Tiger Hill, already at 8,600 feet with one of the best views of the Himalayas for non-climbers. It is possible sometimes to catch a glimpse of Everest but almost certainly Kanchengunga, not far short of Everest, can be seen, caught by the first rays of the rising sun. Or from Calcutta a trip can be made, if permission has been obtained, to the mountain kingdoms of Sikkim, Bhutan and Nepal, and into the attractive state of Assam with its tropical forests and fine wild-life sanctuary at Kaziranga, and to Cherrapunji, claimed to be the rainiest spot in the world.

The foregoing covers only a few of the main tourist attractions of India but even then only the tourist with at least a month to spend on the subcontinent could hope to visit them all and then only briefly. Anyone who knows India will say immediately, why no mention of Sanchi, or Mandu, or Benares and a host of other centres with attractions no less important than, say, the Taj Mahal? The answer is that it takes years to explore this country thoroughly. When I lived in India myself I motored extensively but even so I was taken to task by an Indian newspaper editor who considered my mode of travel superficial. 'There is only one way to see India', he told me, 'and that is by bullock cart.' Even that was more advanced than the method employed by Thomas Coryat. But this first English tourist on his long walk was able to provide a fine description of the Grand Trunk Road, trod in the searing heat of the June sun at the rate of twenty miles a day as 'such a delicate and even tract of ground as I never saw before . . . the most incomparable shew of that kinde that ever my eies survaied'. It was a description which prompted Sir Thomas Roe, Britain's first Ambassador to India and a friend of Coryat, to incorporate the road and the trees which bordered either side in a famous map of India, the earliest British attempt to map the Mughal territories. The road is still bordered with trees.

The great improvement in roads throughout the country in recent years does mean that a car or bus can take you where you want to go. But the deeper you penetrate the more careful must be the planning, and the tougher the constitution. Travel agencies can help but on the whole they have little knowledge of anything off the tourist highroads. There are so many attractive places about which even many Indians are ignorant. The curious traveller always likes to find something that is special and perhaps not visited by others. For example, all those tourists who go to Agra to see the Taj Mahal by-pass much that with a little more time would provide so much more enjoyment. They should stop at Mathura to see its small but fine museum, and, four miles away, go to see Vrindaban, haunt of Lord Krishna. If Buddhist caves are your goal and you are in Bombay, try to see the caves at Karla on the way to Poona. From Bombay a drive to Nasik, plus a short climb would bring you to the source of the Godaveri river, a very holy Hindu place of pilgrimage. A stay at Sanchi in a comfortable tourist hotel enables you to take a leisurely walk round the Buddhist *stupas*, and visit an interesting museum six miles away at Vidisha. I would never advise tourists to visit India without seeing Mandu, the 'City of Joy' on the crest of the Vindhya hills, about 62 miles from Indore. Although now a deserted summer capital, history exudes from every stone, and the surrounding countryside is lovely. When in Mysore try to see the Jog Falls. A visit to Lucknow can also be rewarding for its nostalgic historical associations. Benares, an old city on the banks of the holy Ganges, the home of famous silk and favourite burial (or more correctly burning) ground for Hindus who have cleansed themselves of their sins and seek release from the cycle of rebirth has its, maybe morbid, fascination for non-Hindus; while Chandigarh, a brand new town and now the capital of the Punjab, is remarkable for its modern architecture designed by the French architect Le Corbusier and a British team. And if you are in this region why not ascend on a small mountain railway to Simla in the foothills of the Himalayas? Or even penetrate the Kulu Valley if you are fond of trekking, or hunting, or fishing.

For a restful sojourn, a visit to the small fishing villages on the coast just south of Bombay where there are vast sandy beaches which you can enjoy all to yourself is an experience to be recommended. Finding accommodation will not be easy though. Then there is Mount Abu in Rajasthan, where you can climb to the highest point in India between the Himalayas in the north and the Nilgiris in the south, which has fine Jain temples, containing intricate marble carvings, and is splendid walking country.

There are no real rules to lay down for travel in India. Use commonsense, accept the difference of the subcontinent, and plan accordingly. The official tourist offices throughout India are most helpful and patient. They also hand out most generously extensive and sometimes attractive brochures. The guide books on India can also help. And motorists can seek advice and obtain good road plans from the motoring organisations. You need no extra-large wardrobe, in fact, you can keep your suitcases relatively empty when you leave home and have clothes made for you quickly and cheaply in India. Warm clothes are necessary only in winter or at the higher altitudes. Those using dak bungalows, former post-bungalows set at intervals along the main roads used in the days of horse traffic, will find then very cheap but low on facilities. More expensive but still cheap, and more comfortable, are the Circuit Houses or Government Guest Houses now open to tourists. Travel at night on the roads is sometimes said to hold its dangers; I have never found this but there were dacoits (highwaymen) in central India. Wild animals generally scatter at your approach, but if you carry much in the way of jewellery, then the wild men might be attracted. It is much the same the whole world over. Railway travel in India is cheap and air travel in India in recent years has been much improved with the acquisition of new aircraft.

Remember that the monsoon period, June to September for the Bombay region, although sometimes only intermittently, and a lesser period elsewhere in India, while bringing succour to the earth also means the torrential rains sometimes disrupt communications. The heat at times can be terrific

15

but for the sun-starved Europeans this might even be an attraction.

Commonsense when eating and drinking is to be recommended but overdue attention can detract from your enjoyment. Remember that India observes some kinds of prohibition for drinkers of alcohol in a few parts of the country. A tourist when he arrives can obtain a liquor permit valid for all the country but only necessary where prohibition is in force. This can be a bore, but India is a different country and this is one of the differences you must take into account when visiting it. Drinking is sometimes done in bedrooms of hotels as public drinking is not always permitted.

If you want to be treated as a maharajah, as Air India advertisements claim, then you can still be, even though the heyday of the maharajahs is past. Political conditions in India go through the normal changes and this sometimes affects tourists, for certain areas, sensitive at times, might be banned to foreigners. But generally speaking foreign tourists are welcome everywhere, and the curious ones even more so.

Historical Background

Khushwant Singh

Examine a map of the world and gauge the size of India. Geographically, it is the seventh largest country in the world and, next to China, the second largest in Asia. It covers an area of 1,127,000 square miles stretching over 2,000 miles from north to south and 1,700 miles from east to west. It has land frontiers with Pakistan, Russia, China, Bangladesh and Burma of over 8,200 miles across deserts, mountains and tropical forests. Its coast is over 3,500 miles long.

India can be divided into three major zones – Himalayan, Indo-Gangetic plain and the Deccan plateau. Before Pakistan was carved out of it, the subcontinent had a kind of geographical unity with the land frontiers of the north and the west. The chief importance of these mountains was their impassability – they are the highest in the world, with Everest rising to 29,028 feet and most of them being snow-bound for twelve months of the year, which accounts for the name *Hima* (snow) *laya* (the abode). Historically more important were the few accessible passes which proved regular inlets for invaders. Such were the passes of Bolan, Khurram and Khyber in the north-west and others which linked India with Tibet. They were known to nomad mountain tribes who fed their flocks in the valleys, to tradesmen who brought caravans of merchandise through them and to marauders who used them to invade and loot the rich plains of India. The history of India is a monotonous and tragic repetition of the pattern of such invasions. The invaders crossed over in the autumn before snowfall and descended to the Indian plains just as the ears of wheat had begun to sprout and the air was cool and fragrant. Most battles against the invaders were fought in the Punjab – and if the latter were victorious, which they often were – they spent the winter months systematically looting the rich cities –

Delhi, Agra, Mathura and others – and raiding the winter crops. Then, retracing their steps, they disappeared into the mountain passes through which they had come.

The Himalayas gave Indians the illusion of being guarded by an impassable wall, and the Indo-Gangetic plains gave the illusion of owning an inexhaustible granary. They are an enormous stretch, 770,000 square miles, and are one of the world's longest alluvium tracts, littered with cities, towns, and villages.

The third great division is the Deccan plateau. The hill ranges of the Vindhya and Satpuras and the Narbada and Tapti rivers mark it off from the north. The region is like a vast triangle, with the low-lying ghats running along the sea-coast to form the other two sides. The region has geographical unity but is inhabited by people of two racial groups: the northern half speaking languages akin to the languages spoken in north India and describing themselves as Deccani, and the southern half by Dravidians speaking Dravidian languages: Telugu, Tamil and Malayalam. Indians talk of these three parts of India as the head, the torso and the groins-and-legs of of the one entity that is India. They visualise it as Mother India with her head in the snowy Himalayas, her arms stretched from the Punjab to Assam, her ample bosom and middle (the Indian concept of feminine beauty requires a woman to be well stacked in the bosom and the buttocks) resting on the Indo-Gangetic plains and the Deccan, and her feet bathed by the waters of the Indian Ocean with Ceylon as a lotus petalled foot-stool. In 1947, the Indian subcontinent had its eastern and western extremes lopped off to make the two wings of Pakistan, one of which is now Bangladesh. Mother India has assumed the shape of the Venus de Milo.

Climate and the Seasons

India has a wide range of climate. Most of the Himalayas are perpetually snow-bound with sub-freezing temperatures. Other regions, such as the deserts of Rajasthan, have summer temperatures rising to 125°F in the shade. The climate of the Deccan is, however, temperate.

18

The monsoon is the chief factor in India's climate. A short winter is followed by a prolonged downpour lasting from June to the end of September. This summer monsoon gives mountains their snow, floods the rivers, and inundates the plains, often drowning thousands of villages; it also irrigates the crops. Nor does it fall evenly – some areas are dowsed – Assam has the highest rainfall in the world; others like some districts in Saurashtra, seldom get more than three inches and are, as a consequence, lifeless.

India, instead of four, has five seasons: summer, monsoon, autumn, winter and spring. Summer and monsoon account for more than six months of the year. During these months when searing heat is followed by torrential rains, the Indian can do little, beside find shade or shelter. The monsoon brings in its wake epidemics of malaria, typhus, typhoid, cholera and innumerable other diseases which further enervates the people. And if the monsoons fail, crops fail and there is famine. This is yet another factor adding to the Indian's fatalistic resignation.

Minerals

India is rich in its mineral resources. She has amongst the world's largest deposits of coal – not of very good quality, nevertheless in great abundance. After the United States and France, she has the world's third largest deposits of iron ore. New springs of petroleum are now being tapped and currently pumping out millions of gallons of oil. India produces three-quarters of the world's mica and, next to Russia, the world's largest supply of manganese. India is not so rich in gold, silver, lead, chromite or copper; but she does have the all-too-valuable uranium and also varieties of rare earths.

The People

The population of India today is over 548 millions and is comprised of a variety of races. Although the dividing lines between the different races are never clear, at least six different types can be identified in different regions.

The oldest are the *adi* (first) *vasis* (settlers), the aboriginals of India. They are somewhat negroid in their features and till recently lived in tribes in jungles and mountains, in a boomerang-shaped area extending from Assam in the north-east down to Cape Comorin.

The next large ethnic group is described as Dravidian. They inhabit the southern half of the Deccan plateau. It is assumed that it was the Dravidians who first drove out the aboriginals from the fertile plains into the forests and mountains – as they themselves were later driven southwards by the tribes of Aryans who came from somewhere in Central Asia.

The Aryans inhabit the Indo-Gangetic valley. They were originally of Caucasian stock – blond, blue-eyed – and spoke the same language as their European kinsmen. To this day Greek and Sanskrit have many words in common and the same grammatical structure. Since the Aryans dominated India from about 2,000 BC onwards, India was known as *Aryavarta* – the land of the Aryans – or *Bharat-varsha*, the land of the *Bharatas*, an Aryan tribe which settled between the Indus and the Ganges.

After the Aryans came the Arab Semites, the Mongols and the Europeans. The Semites and the Europeans mingled more rapidly with the local population and do not occupy any particular region. So did the Mongols, but in certain outlying regions of north-eastern India, Mongloid strains are still discernible.

Languages

As one may well expect, such a large number of races produce a large number of languages. Linguists have listed 845 dialects and 225 distinct languages. The Constitution of India recognises 15 major languages including Sanskrit and English. The most widely spoken language is Hindi – or Hindustani. Almost forty per cent of the population speak or understand some form of Hindi which has therefore been recognised and developed as the national language.

Most Indian languages fall into two broad categories: those

of Sankrit paternity spoken in the northern half of India and those of Dravidian origin spoken in the south. The northern half, though much injected with Arabic and Persian, share a large vocabulary in common; the southern half though liberally infused with Sanskrit terminology, are distinctly different from the northern languages and have much in common between them. India has also produced its own brand of English, known as the Hobson Jobson, which has a sizeable dictionary of its own.

The discovery of the evidence of India's earliest civilisation is itself a story. Some time in the middle of the nineteenth century, British engineers were engaged in linking the port of Karachi with the cities of the Punjab. The task of making solid foundations for laying the rail-track over the hundreds of miles of desert was given to two brothers, locals of the region who had heard that enormous quantities of sun-baked bricks were theirs for the digging. They dug these up and laid the track, at enormous profit to themselves. It was later discovered that these bricks were more than 4,000 years old.

In the 1920s the archaeological department of the Government of India decided to look into this story. As a result two buried cities were excavated – Mohenjodaro (mound of the dead) along the Indus, and Harappa some 400 miles away on the Ravi. Thereafter many other excavations in the Punjab (notably Rupar) and Lothal in Gujarat yielded a bumper harvest of seals and terracotta, unfolding a tale of many cities that flourished some time before 3,000 BC. They were in a broad belt about 500 miles on each side of the River Indus and 1,000 miles along its course. Some cities, like Mohenjodaro, must have existed for some centuries, as nine layers of construction were unearthed. Since time and trade and geography seemed to link these cities, the period in which they existed became known as the Indus Valley civilisation, contemporary with the Mesopotamian and the Nile civilisations.

A great deal of literature has been produced on the Indus Valley civilisation. Earthenware and pottery, metal jewellery, garments of wool and cotton, figurines of kings, warriors, dancing girls, trees, animals (real and imaginery), god and goddesses and toys can be seen in the museums. Inferences about the organized city life have been drawn from the exis-

21

tence of broad stronger roads, water-ways and sewage systems, public baths and meeting places. The existence of funerary mounds – some with skeletal remains which indicate sudden death by violence or natural catastrophe, point to the way these cities were destroyed. Amongst the finds in these cities were seals with pictographs which have not yet* yielded up their secrets. If and when the words are deciphered we may have a major break-through into our past history. Till then, we accept what scholars like Sir Mortimer Wheeler and Basham tell us about the Indus Valley civilisation – that its halcyon years were between 3,000 and 1,500 BC, that the people who created it, for want of a better alternative, be described as Dravidians,** and that they were overcome by the ruffianly Aryans mounted on horses and riding in chariots.

The Aryans were Caucasians consisting of pastoral tribes who lived in villages (*grama*) under an elected chieftain (*raja*) who was advised by a council of elders (*sabba*) – and on cases touching the interests of the entire village by the *samiti* – meeting of the village. They brought a rich language, Sanskrit, with them. They developed this in India, and the *Vedas*, *Upanishads* and the *Gita* are only one aspect of Sanskrit literature. They achieved an equally high standard in drama and poetry.

By about 800 BC the Aryans learnt how to smelt iron ore for tools and weapons. Thus armed, they were able to forge ahead further east and southwards. It is likely that the eastward shift was due to the discovery of large deposits of iron ore in Bihar and the development of inland river traffic and overseas trade. The centre of power shifted from the Punjab to Uttar Pradesh, Bihar and Orissa. New cities rose into prominence: Indraprastha (present-day Delhi), Hastinapur, Ayodhya, Kashi (Benares) and Pataliputra (Patna). As the centre of Aryan power shifted eastwards, their religious system – Brahmanical Hinduism – and protest movements like Jainism and Buddhism, began in these eastern regions. At the same time the Aryans neglected the defence of their north-

*Early in 1969 some Scandinavian and, later, Indian scholars claimed to have deciphered the writing on the seals.
**Different racial types ranging from the Latin Mediterranean to the Polynesian Australoid are depicted amongst the human figures. The formation of skulls reveals the same ethnic diversity.

west frontiers, thus leaving them vulnerable to invasion. The first invaders were Persians under Cyrus, followed by Darius (521–485 BC). Darius conquered the Indus Valley and apparently recruited Punjabi soldiers who fought with Darius' son Xerxes in the Persian invasion of Greece (479 BC). Although this Persian invasion was a passing phenomenon it left a permanent impress on religion, art and administration. The solar cult of the Hindus is a Persian import. The use of pillars to popularise laws is also said to have derived from the Persians, as is the notion of divine monarchy to strengthen the rulers' position in society.

The next foreign intruders were the Greeks. Alexander of Macedon overthrew Darius III in 331 BC, and in the winter of 328–327 BC advanced into India. He was welcomed at Taxila and sumptuously entertained. Alexander proceeded further inland to the Beas and defeated King Porus and his vast army which included hundreds of war elephants. He was not able to penetrate much further into the Punjab, but marched a short way along the Indus before returning home to die. This invasion was no more permanent than that of the Persians, but Alexander left behind him scattered colonies of Greeks who influenced the Indian style of art, architecture and sculpture. Greek scribes noted the existence of marriage markets (where poor parents sold their daughters) and the practice of the immolation of widows on the funeral pyres of their dead husbands (*sati*).

The Mauryas

Indians soon reacted to the Greek invasion. A powerful Aryan dynasty, the Mauryas of Magadha (present-day Bihar) rose under Chandragupta (a popular name amongst Indian monarchs), who defeated the Greek general Seleukos Nikator in 305 BC and expanded his domains across the Indo-Gangetic plain from Bengal in the east into the heart of Afghanistan in the north-west. An account of his rule has been left by Megasthenes, Nikator's ambassador in his court.

Chandragupta's twenty-four years reign is notable for a treatise on the art of government, *Arthashastra*, written by his

23

Minister Kautilya Chanakya (321–296 BC). Although Kautilya's thesis outdoes Machiavelli in diabolic amorality, it is more than an essay on the art of overcoming enemies and acquiring power, and contains much useful advice. He made punishment (*dandaniti*) the basis of administration, and his views on diplomatic relations are direct, pragmatic and full of commonsense. He uses concentric circles, *mandalas*, to illustrate his point – one's neighbours are one's natural enemies. One's neighbour's neighbour on the other side is his enemy and therefore one's friend.

Megasthenes' original work was lost, but Pliny quotes his tributes to Indian civilisation: no slavery, peaceful husbandry, industry, chivalry of the men, and virtuous chastity of the women. He mentions the existence of seven classes of Indian society – scholars, tillers, graziers and hunters, artisans and traders, police (including spies) and bureaucrats.

Chandragupta's son, Bindusar (298–232 BC) added the Deccan to his domains. He exchanged envoys with the rulers of Egypt and Syria, and his son Asoka (273–232 BC) extended his sway further south into Mysore. He then went to war to extend his territory across Orissa up to the Bay of Bengal, and a bloody battle was fought at Kalinga (261 BC) giving a pyrrhic victory to Asoka. Overcome with remorse, he renounced violence as an instrument of policy and turned to Buddhism for solace. He had his messages of peace inscribed on rocks and pillars all over his domain, and sent out missionaries to propagate Buddhism. His son Mahendra became a monk and took the gospel of Buddha to Ceylon. An adaptation of Asoka's Sarnath Lion Column has been adopted by the Indian Government as its official emblem, thus perpetuating his memory.

Mauryan power petered out within a hundred years after Asoka's death. New invasions of Bactrian Greeks dealt the *coup de grâce*. During this time Brahmin rulers established themselves in the Indo-Gangetic plains while the Deccan was ruled by the Telugus. Though the 'Lords of the Deccan' were Hindus it was during their rule that Buddhist monks hollowed out caves at Karla and raised *stupas* (burial mounds) at Sanchi and Amaravati. The south was divided into three kingdoms: the Cholas round present-day Madras, the Cheras in Kerala

24

Four faces. Top (l to r): *A Kulu girl from Himachal Pradesh and a child from Kashmir.* Bottom: (l to r): *A Mizo from Assam and an old man from Gujara*

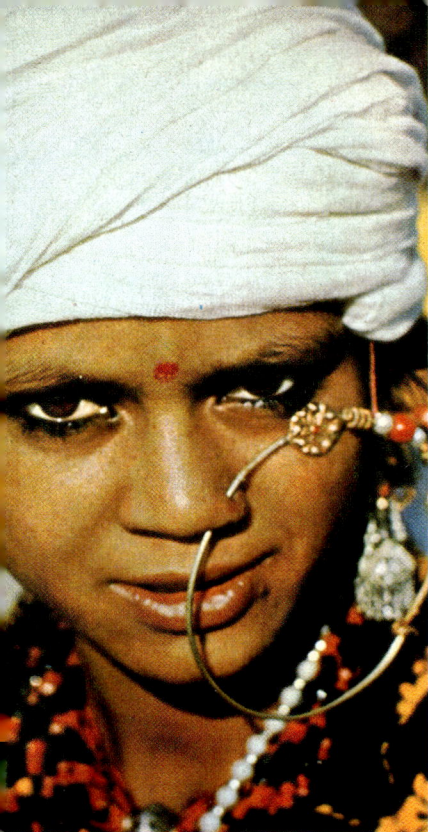

and the Pandyas based in Madurai and reaching down to Kanniyakumari. Madurai became the centre of Tamil learning. These powerful southern dynasties made their influence felt all around. The Tamils invaded Ceylon, traded with the Romans on one side (gold Roman coins have been found in Tamilnad) and the Far East on the other. With sandalwood and spices, teak, ebony, and ivory, Tamilians exported Hinduism, Buddhism, Jainism, and Indian philosophy, art and medicine.

While these changes were taking place, the Greeks again appeared in the north, Menander occupied the Punjab, made Sialkot his capital, and then pushed eastwards. He invaded the Mauryan capital Pataliputra (Patna) in 150 BC. Some Greeks – known as *yavanas* (Ionians) – accepted Indian religions and Menander himself was converted to Buddhism.

Invaders continued to make their inroads through the north-western passes. This time most of them were from western China. After the Greeks came the *Sakas* (Scythians) around 130 BC; then the *Pahlavas* (Parthians), and after them the Kushans from central Asia. The first two consolidated power in the north and then pushed east and southwards. The Kushans first established themselves at Peshawar and then sent their armies along the Ganges as far as Benares. One of the rulers of this dynasty, Kanishka (78–144 AD), gave India a new calandar, beginning in 78 AD. He became a Buddhist and was also a patron of the arts.

There is a hundred-year hiatus between the disintegration of the Kushans (200 AD) and the rise of the greatest of the Hindu dynasty that ruled India, the Guptas.

The Guptas

Four Gupta monarchs, Chandragupta, Samadragupta, Chandragupta II (also known as Vikramaditya) and Kumargupta were based originally in Magadha (Bihar). At the height of their power they had all the Indo-Gangetic plain down to the northern fringes of the Deccan under their control. They and their descendants ruled for over three centuries – from 300 AD to the death of Harsha in 647 AD. This period of Indian

25

A South Indian bangle seller.

history is known as the Golden Age of the Guptas. Peace and stability saw the blossoming of art and literature, and the somewhat crude Pali was replaced by the highly polished and grammatically perfect Sanskrit.

The Advent of Islam

The Muslim conquest of India had the most profound impact on the political, social and cultural life of the country. Islam had come to India before the Muslim conquerors, brought by Arab traders to the western coast within a few years of the death of the Prophet. Then, in 712 AD the seventeen-year-old Mohammed Bin Qasim invaded Sind. This incursion had one result; it showed the Muslims that India was rich and ready for conquest.

It was, however, another two centuries before the Muslims mounted a full-scale invasion of India. This was left to Mahmud of Ghazni. Between 1000 and 1027 AD he invaded India 17 times. He decimated the Hindu opposition, destroying temples and sacking cities along the Jumna and the Ganges. His most spectacular victory was on his sixteenth invasion when he crossed the Sind desert and captured the holy city of Somnath on the Arabian Sea. He destroyed its great temple and carried enormous wealth back with him. Ever since, the name Mahmud has been repugnant to Hindus – and the rebuilding of Somnath attained symbolic significance for them. Mahmud terrorised Hindu India. The historian Ibn Zafir mentions an Indian custom whereby a victorious rajah cut off a finger of his vanquished foe. According to him Mahmud followed this practice and acquired a hoard of severed Hindu fingers. With savagery towards the 'infidel', Mahmud combined sophistication and patronage of arts and literature. One of the greatest poets of the Persian language, Firdausi, lived in his court. So did the historian Al Beruni.

The real conqueror-founder of Islam in India was Mohammed Ghauri who first invaded India in 1175 AD. He came only as far as Multan. His second attempt was thwarted but on his third and fourth invasions he overran most of the Punjab. Then he came up against united Hindu opposition led by the Rajput chief Prithvi Raj Chauhan, and in the battle of Tarain, fought in 1191, the Rajputs defeated the Muslims. The next year Mohammed Ghauri returned and on the very same battlefield defeated and slew Prithvi Raj. Ghauri's viceroy, Qutb-ud-din Aibak, who later succeeded him as Sultan, captured Delhi and Mathura and territories between the Jumna and the Ganges. Thereafter the invaders took on other Rajput chieftains and eliminated them one by one. The ease with which Muslim invaders permeated India is scarcely credible. A Muslim general, Mohammed Khilji, conquered Bihar in 1193, destroying Buddhist *viharas* and massacring Buddhist monks. Six years later, he pushed on into Bengal.

Mohammed Ghauri did not spend very much time in India but his slaves, whom he appointed as viceroys, consolidated the conquest of the country. Thus came about what is known both as the Sultanate and the Slave Dynasty of India. First, there was Qutb-ud-din a generous tyrant, who was followed by Iltutmish (1210–1235); he in turn was succeeded for a brief period of three and a half years by his daughter, Razia Sultana (1236–40). The Slave Dynasty declined after Ghiyasuddin Bulban, who ruled northern India with an iron fist for 40 years, died in 1286. It might be profitable to discuss here why Hindu India crumbled before the Muslim onslaught.

Perhaps the most decisive cause of Hindu defeat was their disunity. Hindus were always eager to see their own enemies destroyed even if they knew that their own destruction would follow as a matter of course. This led to the defeat of Prithvi Raj, then to that of his rival Jai Chand, at the hands of the Ghauris. Even on the field of battle, Hindus refused to have a unified command and fought under their respective chieftains. Desertion by a chief often became a rout. Their caste system restricted their fighting manpower. Only a small proportion of Hindus were trained to use arms; the rest of the population remained spectators, often indifferent to the fate that befell the arrogant warriors. Nor did the Hindu technique of warfare match that of the Muslims. The former used elephants or foot soldiers armed with lances or maces. Muslims rode horses and had archers. The elephants frequently turned back and trampled on the soldiers behind them, while Muslim archers inflicted heavy slaughter from a safe distance. Another

27

factor was size. The Turk, Mongol, Afghan and Persian soldier was physically bigger and stronger than the Hindu. He was as much a crusader fired with fanatical zeal to destroy the infidel as a freebooter intent on looting the wealth of Hindustan.

But to return to the succession of Muslim dynasties. The Slaves were followed by another Turkish dynasty, the Khiljis. By now the ruler had to fight on two fronts, against the Indian resistance on one hand and against Mongol tribesmen who had begun to knock at the north-western gates of India on the other. The most distinguished of the Khilji monarchs was Ala-ud-din, who ruled for twenty years (1296–1316). His general, Malik Kafur, had conquered the Deccan and the whole of south India by 1312 – thus justifying the title his master had assumed for himself *Sikander-i-Sani*, Alexander the Second.

The Khiljis soon gave way to another Turkish dynasty, the Tughluq, who ruled from 1320–1397. Their great king was the eccentric Mohammed Tughluq (1325–51) whose empire was almost as large as Ala-ud-din Khilji's. He experimented with having two capitals, one at Delhi and the other in the Deccan at Daulatabad; introduced token currency – an experiment which drained his treasury of all its gold and silver, and raised new kinds of taxes.

Mohammed's successor Firuz, (1351–88), was also a great builder of cities and waterways as well as a patron of learning. The Tughluq decline began during his rule. The last effective ruler of this dynasty, Nasir-ud-din, was unable to withstand the onslaught of the Mongol tribes. In 1397 Taimur (Tamerlane to the European) crossed the Indus into India. 'My object in the invasion of Hindustan is to lead an expedition against the infidel,' he recorded, 'to purify the land from the filth of infidelity and polytheism; and that we may overthrow their temples and idols and become *Ghazis* (slayers of Infidels) and *Mujahids* (warriors of Islam) before Allah.' He more than justified the title of *Ghazi*. Before engaging the armies of the Sultan of Delhi, Taimur ordered the massacre of 100,000 Hindu prisoners taken in the Punjab campaign. Thus 'the sword of Islam was washed in the blood of infidels' and many Hindus terrorised into accepting Islam.

No one could withstand Taimur. After plundering the Punjab he ransacked Delhi. The immediate effect of his invasion was to break India into two; the northern half returned to the control of the Turks and Afghan princes; the south again became independent under Hindu kings. But this was only an interlude. The Saiyyads (1414–47) and the Lodhis (1447–1526), both from Afghanistan, ruled parts of the north while the Mongols were getting ready to re-enter India.

As Muslim power weakened, Hindus wreaked vengeance on the Muslims. The majority of early Muslims were in fact Turks, Afghans, Persians and Arabs who came with the invading armies, found Indian wives and made their homes in India. From the nucleus of these communities, Islam spread amongst Hindu tribes – sometimes due to political influence, more often through the proselytising of the Mullahs and the Sufi mystics. Most of the converts were Hindus of the lower castes who could no longer turn to Buddhism: there were no Buddhists left to turn to – so their only escape from the caste discriminations of Brahmanical Hinduism was Islam.

The Mughals

No other dynasty has left as great a mark on India as the Mughals. A succession of six monarchs ruled the country for two centuries before their progeny dissipated their inheritance, suffered humiliation at the hands of new conquerors and their erstwhile subjects, and finally bowed out to make way for the British.

The first Mughal monarch, who ruled from 1526 to 1530, was Babur. As a descendant of Changez, Hulagu and Taimur, Babur believed he had a duty to conquer India. His ambition was fired by stories of India's wealth and as he occupied Kabul, as he later wrote in his memoirs he 'never ceased to think of the conquest of Hindustan'. In the course of the seven years from 1519 to 1526 he entered India five times. 'Do not hurt or harm the flocks or herds of these people, not even to their cotton ends or broken needles,' he told his soldiers. 'Possession of this country by a Turk has come down from of old; beware not to bring ruin on its people by giving way to fear or anxiety; our eye is on this land and on this people;

29

raid and rapine shall not be.' This was said in 1519 on his second foray, when he had occupied much of north-western Punjab up to Bhera and Khushab.

Babur the next year returned and learnt that disputes between different branches of the Lodhi Afghans (one of whom was in possession of Delhi) could be exploited to his advantage. After a pause of four years Babur extended his conquests in the Punjab and made preparations to take Delhi. And finally, on November 17, 1525, he 'set out on my march to invade Hindustan...putting my foot in the stirrup of resolution, and taking in my hands the reins of faith, I marched against Sultan Ibrahim, son of Sultan Sikandar, son of Sultan Bahlol Lodhi Afghan, in whose possession the city of Delhi and Kingdom of Hindustan at that time were.'

Babur chose an easy victim. Ibrahim Lodhi had lost the support of his Afghan kinsmen, and although he was able to muster a host of 100,000 men to oppose the 10,000 under Babur's command, not many had the stomach for battle. The issue was settled on the morning of April 21 1526, on the field of Panipat. Babur's superior tactics won him a speedy victory in six brief hours. By noon 20,000 of the Lodhis including Sultan Ibrahim were dead and the rest had fled. Babur did not give the vanquished time to recover. He sent his son Humayun on to Agra, while he himself marched to occupy Delhi. The capture of Delhi and Agra (Humayun obtained the fabled *Koh-i-noor* diamond at Agra) did not bring the rest of northern India to heel. Babur had exhorted the Indians not to 'fly from our conquests'. Many returned to their possessions, and some, like the Rajputs, banded together to expel him from the country.

On March 16, 1527, a very nervous Babur, forswearing wine and calling upon Allah to help him in a holy war, met the Rajput host at Kanhua, thirty miles from Agra. The Rajputs were led by Rana Sanga whose body bore marks of his martial exploits: he had lost an arm and an eye and had eighty scars on his person. Once more God granted victory to Babur, who expressed his gratitude to Allah by raising a pyramid of pagan skulls and investing himself with the title of *Ghazi* – slayer of infidels. Thereafter Babur and his sons took other provinces piecemeal. He had come to stay, but liked neither

30

India nor the Indians. 'Three things oppressed us in India,' he wrote, 'heat, violent winds, dust.' And in another memorable passage, he denounced almost all he saw of the Indians. '...a country naturally full of charm but its inhabitants are devoid of grace. They have no genius, no comprehension, no politeness, no generosity, no robustness of feeling. In their ideas, as in their methods of production, they lack method, art, rules and theory. There are no baths, candles, torches, schools or even candle-sticks.' All he found to praise was India's countryside, its monsoon and its gold. Having savoured the first two and acquired much of the third, he longed for the cool, fragrant air of his homeland and the melons of Kabul. According to legend he brought death upon himself. When his elder son was taken ill and pronounced beyond cure, Babur walked round his son's bed three times repeating 'O God! If a life be exchanged for a life, I who am Babur give my life and my being for Humayun...I have borne it away.' And so it was. Humayun recovered and Babur was stricken. 'Lord, I am here for thee,' were the last words he said after expressing his wish to be buried in Kabul.

Humayun was 23 years old when he inherited Babur's kingdom. He was temperamentally unable to cope with all the problems his doting father had left him. He was a scholar and enjoyed Turkish, Persian, Arabic and, later, Hindu literature. He studied mathematics, philosophy, astronomy and astrology. He wanted to be left alone. Babur had arranged to have him trained in administration – an art he himself had neglected – and had nominated him Governor of Badakshan and later of districts of the Punjab. He was given military training. But it was obvious that Humayun could not make up his mind between sovereignty and solitude. His father loaded the scales against him. Although he recognised Humayun as his heir, he had given his other three sons large tracts which in effect divided and weakened the kingdom. He had not had the time to destroy the Afghans who quickly reasserted their power in different regions While Humayun was dallying at Agra or building a suburb of Delhi, Afghan chieftains, notably Sher Shah Sur, began to rally in Bihar and Bengal. Humayun moved against him. Sher Shah feigned retreat, then turned round and cut off Humayun's contact with his capital – where meanwhile one of Humayun's brothers had proclaimed himself Emperor.

Sher Shah Sur awaited the monsoon. Then on June 26, 1539, at Chausa, he attacked Humayun's water-logged army and slew eight thousand Mughals. Humayun escaped but Sher Shah Sur pursued him all the way to Agra and into the Punjab. Threatened by his brothers, Humayun fled through the deserts of Sind and out of Hindustan. For the next five years until his death in 1545, Sher Shah Sur was Emperor of India. Few men have left so lasting an impression of their administrative genius as did this Afghan commoner who had become king. His able minister Todar Mal organised the revenue system of India which survived until British rule. Sher Shah built roads to link the major cities of northern India, built rest-houses for travellers, and abolished tolls. He raised his Hindu employees to positions of eminence and recruited Rajputs to his army. During his five-year reign there was peace in the land, and no-one knew or cared about the fate of Humayun.

Sher Shah's successors – first his son Islam Shah (1545–53) and, on his death, his son Firoz (murdered three days after his accession) and finally Sher Shah's brother Adil Shah were unable to keep the kingdom together. By 1556 there were four independent Afghan kingdoms and the most powerful man was a Hindu shopkeeper, Hemu, who ruled Delhi in the name of his Suri patron. The latter years witnessed the failure of monsoons followed by widespread famine and epidemics.

Aided by the Persians Humayun was now able to return to India. Once again the decisive battle was on the field of Panipat. On November 5, 1556, Humayun clashed with Hemu's vast army of 1,500 war elephants and an assortment of horse and foot soldiers. The elephants panicked; Hemu, hit in the eye by a stray arrow, was deserted by his men. Humayun quickly proceeded to reoccupy Delhi and Agra. A few weeks later, while studying in his tower library in Delhi, he heard the call for prayer from a minaret of a neighbouring mosque built by Sher Shah Sur. Humayun hurried down, and missing his step, fell and fractured his skull. Thus ended the short and interrupted rule of one described by the historian T.G.P. Spear as 'the problem child of the Mughals'.

Humayun's son, Jalaluddin Akbar was a rootless child, born of fugitive parents. He was nursed by a foster mother and denied any chance of a formal education in Persian or Arabic, but he inherited a sharp mind which enabled him to memorise works of poets like Hafiz and Rumi. When his father died, Akbar was only fourteen. He was proclaimed Emperor of Hindustan but power was wielded on his behalf by a kinsman, Bairam Kahn, a ruthless man who soon became drunk with power. It took young Akbar four years to get rid of him. For another two years the state was administered by his foster mother and her son Adham Khan. Akbar spent these two years hunting tigers and elephants, but when Adham Khan roused his ire, first by his greed and then by murdering a senior minister whom he suspected of being a rival, Akbar had Adham Khan hurled from the ramparts of the fort at Delhi and finally became Emperor of Hindustan in fact as well as name.

The year 1562, when Akbar was twenty, marks an important point in his career. That year he not only came into his own inheritance, he also launched an experiment to unite his subjects, both Muslim and Hindu, and set an example by himself taking a Hindu wife and raising her Rajput kinsmen to positions of eminence. This brought him into personal contact with the Hindu faith. The following year he ordered the abolition of the tax on Hindus going to their places of pilgrimage and followed it up by abolishing the tax paid by them in lieu of military service. Akbar then proceeded to bring the whole country under his dominion; as he put it, 'A monarch should be ever intent on conquest, otherwise his neighbours rise in arms against him'.

The title of 'Great' attached by historians to Akbar's name was not by virtue of his conquests but for his humane administration, ('If I were guilty of an unjust act,' he once said, 'I would rise in judgement against myself'). His reforms in revenue matters and the civil services, his patronage of the arts and architecture, and above all his concept of a secular state in which his subjects of different races and religions enjoyed equal rights, and his attempt to formulate an eclectic faith combining what he considered the best of Islam, Zoroastrianism, Hinduism, Jainism and Christianity; with him-

33

self as the divine god-head. He was fortunate in his advisers. There was Abdur Rahim Khan-i-Khanan, son of Akbar's guardian-tutor of the earlier days, Bairam Khan, and Todar Mal who had done such an excellent job for Sher Shah Sur, as well as Birbal, another Hindu who combined an uncanny ability for accountancy and administration. Akbar also had his coterie of sycophants. Amongst them was the talented family of scholar-philosophers Shaikh Mubarek and his sons Abul Fazal and Faizi. There were a few, like the historian Badaoni, who hated Akbar's trifling with Islam and courtiers who encouraged un-Islamic innovations, but he could do little except turn his venomous quill to writing a private diary.

Akbar's father and grandfather had not had much time to patronise the arts, and it was left to the uneducated Akbar to make amends. He collected a library of 24,000 manuscripts, and surrounded himself with scholars, poets, theologians, painters and musicians. It was during Akbar's reign that India had its first real contact with European powers, notably the Portuguese and the English. In 1603 John Mildenhall paid a courtesy call on the Emperor at Lahore and presented a letter from Queen Elizabeth. The Portuguese Jesuits, already in India, were much put out by the arrival of 'English thieves and spies'. All Akbar was able to do was to ensure that pilgrims and sea-going trade were not interfered with. His last years were saddened by the rebellion of his son and heir-apparent, Jehangir. He kept the prince under house arrest for four years and considered disinheriting him in favour of a younger son, but death overtook him before he could accomplish this, on October 17, 1605.

Jehangir was as different from his father as he could be. He was more erudite but far less magnanimous; and although born of a Hindu mother and married to a Hindu princess, (who killed herself when shamed by her husband's revolt against his father), he had none of his father's vision towards a united people. As a result his many achievements are overlooked. He kept his kingdom together by keeping down the Rajputs and quelling attempts at insurrection by Deccani Muslims and he did make notable advances in the art of building. His father's mausoleum at Sikandra (seven miles north of Agra), the bejewelled Itmad-ud-Daulah on the Jumna, on which marble craftsmen showed superb skill in stone inlay work, the Royal Mosque and the gardens in Lahore, and in the Vale of Kashmir – all bear testimony to his taste. Nevertheless, in the popular mind, Jehangir's twenty-two-year reign is remembered for his revelry, bouts of drinking, dalliance with women, and his sadistic delight in ordering and watching executions. He failed, however, to keep down his ambitious and impatient progeny.

Shah Jahan, like his father, was 35 when he became king and like his father had to destroy his brothers before he was able to secure the throne for himself. Fratricidal contests for the throne had by now become an established tradition of the Mughals – as a result, the many principalities which had either sided with the loser or made a bid for freedom had to be reconquered. Shah Jahan spent sixteen of his thirty-two years as Emperor campaigning against the Rajputs and the Deccan Kingdoms. He, too, although he had a generous dose of Hindu blood in his veins, victimised his Hindu subjects and even destroyed their places of worship. His shortcomings as a monarch, however, were obliterated by the great works of art executed in his time. The palaces and the Pearl Mosque in the fort of Agra, the Royal Mosque – the biggest and the most beautiful of all mosques of the world – in Delhi, the Red Fort in Delhi, and innumerable gardens, culminating in the divinely conceived and executed Taj Mahal, raised to enshrine the dust of his favourite queen and mother of fourteen of his children. Shah Jahan spent the last seven-and-a-half years of his life under house arrest, ordered by his son Aurangzeb, and died in January 1660.

Historians are sharply divided in their opinions of Aurangzeb. Nor is the controversy limited to history books, it is a live issue amongst the historians of Pakistan on one side, and Indians on the other. Aurangzeb ruled an empire bigger than Akbar's and for as long. He was a puritan, and did not touch liquor. He inherited few of the amorous traits which had sullied the names of his forefathers; a devout Muslim, he spent hours in prayer and in making copies of the Koran. He dressed simply and lived frugally. He was fearless and was constantly putting down rebellions and extending his domains: 'A ruler should always be on the move... being in one place gives the impression of repose but brings a thousand calamities,' he wrote. He was also calculating, ambitious and treacherous. He tricked and slew his brothers and their sons; he had kept his father in imprisonment; and reduced his non-Muslim subjects to the rank of second-class citizens. When they revolted against him, he put them down with a severity which renewed the image of the Muslim invader as a bigot and a savage. So great a stickler to the letter of Islamic law was he that despite his learning, he looked upon the arts and music as evil indulgences. Descended as he was from a long line of builders, all he left behind him were mosques (a very beautiful one in the Red Fort in Delhi), and another in Benares raised over the ruins of the Hindu temples he destroyed. In short, he was a Jekyll-and-Hyde character.

Historical justice ordained that the very people that Aurangzeb sought to destroy themselves destroyed the Mughal rule in India – the Jats, the Sikhs and above all the Marathas, who rose under their great leader Shivaji, and finally reduced Aurangzeb's successors to the position of pensioners. Aurangzeb's son Bahadur Shah spent much of his energy fighting his own brothers, and then the Sikhs who had risen under Banda. He died broken-hearted and almost insane with frustration.

Thereafter one Mughal followed another in quick succession – brother killed brother, uncle killed nephew. Some gave themselves up to drink, others to common whores or to homosexual orgies. The most colourful of these debauchees was Mohammad Shah, aptly named *Rangeela*, the colourful one. His domain was now confined to the city of Delhi. India was wide open to the invader.

The first powerful blow was struck by the Persian invader Nadir Shah. He invaded India in 1739, defeated Rangeela's forces at Karnal and occupied Delhi. He stripped the capital of all its wealth. On one morning in March, he ordered a general massacre of the citizens and slew 150,000 men and women. In five months' stay in India he destroyed her system of administration, and helped himself to her treasure, including the fabulous Peacock Throne and the Koh-i-noor diamond. All he left was the Mughal King and the shattered shell of what at one time had been the most powerful empire in

Asia. The final blow was struck by Nadir Shah's general, Ahmed Shah Abdali. In a series of nine invasions, Abdali completed the task of destruction. On his fifth invasion, in 1761, he defeated the Marathas at Panipat, thus giving the Sikhs a chance to fill the vacuum in northern India. The Marathas recovered from the blow, and thereafter he had to contend with the Marathas, the Jats, the Sikhs and the English for power. The Mughal king was a monarch only in name. His territory extended from Delhi to Palam – seven miles west of the capital:

Saltnat Shah Alam
Az Dilli ta Palam

What were the major achievements of the Mughul rule and the chief causes of its downfall? These two are standard questions for all history examinations in Indian universities – and answered in the standard six or seven answers. Achievements? Two hundred years of security from external invasion and internal disorder. Two hundred years which gave northern India political unity. Closer contacts between Hindus and Muslims. Development of the Urdu language and etiquette. Development of painting, music, architecture and historiography. Contact with European powers. And so on.

Causes of downfall? Decadence of the monarchy and nobility leading to the demoralisation of the army and the civil service. Insufficient attention to agriculture and commerce, leading to economic bankruptcy. The religious intolerance of Shah Jahan and Aurangzeb, leading to Hindu resistance movements among the Rajputs, Marathas, Jats and the Sikhs. Making the privileged Muslims indolent and parasitical. Weakening of the central government, wasteful campaigns in the Deccan and consequent disability to resist foreign invasions, notably that of Nadir Shah in 1739, followed by encroachments by the British.

Before and After Independence

Frank Moraes

My memories of British rule in India go back to the First
World War, dim memories of that time for I was seven when
war broke out. In school we saw India largely through British
eyes, and I suppose if we were asked to list the landmarks in
Indian history, most of us would have cited Plassey, the Black
Hole, and the Mutiny. India as India was part of a fading
palimpsest.

Plassey (1757) might have signalled the establishment of
British rule in India, but as a battle it was ridiculous. The
victors suffered 65 casualities and even the toll of the defeated
was under 500. Plassey was followed by the first shaking of
the pagoda tree which brought Clive his notorious *jagir*,
initially consisting of the quit rent of the Twenty-four Par-
ganas.

In the eyes of contemporaneous India the battle of Plassey
carried no political significance. More important to the
various protagonists was the Anglo-French struggle for
supremacy in the country which ended with the surrender of
Pondicherry by the Comte de Lally in January 1761. Lally,
a bold but impetuous soldier, had been defeated just a year
earlier by Sir Eyre Coote at Wandewash. After Aurangzeb's
death in 1707 the Mughal empire continued as a phantom
dynasty until 1858 but it must have been clear even to Clive's
contemporaries between the four years, 1757 to 1761, that
the empire was in its last throes. Lally's surrender of Pondicherry marked the end of French power in India. The French
dream of empire was over, though lingering remnants of
French influence survived in a few pockets.

The decay of the Mughal empire had induced a non-
Muslim renascence which Aurangzeb had earlier accelerated
by his religious intolerance. By 1758 the Maratha power was

38

at its zenith, and the chiefs of the Scindia and Holkar families had advanced as far as Delhi where after Nadir Shah's death (1747) the Afghans had replaced the authority of the Persians. In this period until 1785 central and northern India were entirely in Indian hands, and in 1761 it seemed as if the prize of Hindustan lay at the feet of either the Afghans or the Marathas. On January 14, 1761, the Marathas and Afghans collided on the historic field of Panipat where the fate of northern India if not of Hindustan had been often decided previously. Panipat was more than a defeat for the Marathas. It was a catastrophe for their cause, for though the resilient Marathas were outwardly to recover themselves, and even to cross the Chambal again by 1767, their old *condottiere* spirit was to assert itself in the development of separate territorial fiefs with the rise of the Maratha succession states. Scindia set up his domain in Gwalior, the Gaekwars in Baroda, Bhonsle in Nagpur and Holkar in Indore.

Along with the Marathas there had come into being the Sikhs whose militancy owed much to Aurangzeb's persecution. Between 1685 and 1708 the Sikhs under their tenth guru, Gobind Singh, developed into a military clan not unlike the knights templars of medieval Europe. Later the Sikhs, under the great Ranjit Singh, were to consolidate themselves into a powerful military state west of the Sutlej. The Marathas and the Sikhs were the last two Indian races to go down before British arms.

At Panipat it wasn't only the Marathas who were defeated. Though not militarily the victors, the Afghans were also compelled to retreat to their hills for the ignominious reason that Mughal Delhi could not provide them with pay. Oddly enough, the third side of this triangle, the Mughals, were not unduly rocked on their thrones; a large part of India accepted them as legitimate and ordained rulers. As a British historian has put it, 'the courtiers bowed and the trumpets blew.'

But not for long. The Mughals were on their way out. Logically the successors of the Mughals should, despite their freebooting traditions, have been the Marathas but their defeat by the Afghans on the field of Panipat followed by the virtual evaporation of the victors gave the Mughals a further indeterminate lease of life.

By 1817 Maratha resistance was virtually at an end following the defeat of the last Peshwa, Baji Rao II, on the plain of Kirkee, near Poona. There followed the disastrous battles of Koregaon and Ashti. With Baji Rao's defeat the British controlled the whole of India east of the Sutlej. A legendary resistance hero of the decade preceding the establishment of British rule in the greater part of India was the Muslim ruler of Mysore, Tipu Sultan, who died fighting in Seringapatam in May 1799. As the price of continued existence many Indian princes of that time accepted British suzerainty. They included most of the Rajput princes, fragmentary relics of the Mughal empire like the Nizam of Hyderabad and survivors of the Maratha confederacy like Holkar and Scindia. The British were now the paramount power in India.

Only the Sikhs, west of the Sutlej, continued to defy and resist the inroads of Britain into India. Ranjit Singh, the greatest of the Sikh leaders, had no love for the English but was wise enough to temporise and in 1809 signed a 'treaty of amity' recognising the Sutlej as the boundary between the British and the Sikhs. Ranjit Singh died thirty years later. About this time the British suffered a severe reverse at the hands of the Afghans in Kabul, and emboldened by this development the Sikhs, unmindful of their treaty with Britain and free of the restraining hand of Ranjit Singh, rashly decided to attack Delhi. The result was calamitous.

Crossing the Sutlej in December 1845 they engaged in a series of fierce battles but were badly mauled and defeated. Young Dhuleep Singh was installed as Maharaja by the British at Lahore where Sir Henry Lawrence was set up as Resident. With him were associated a group of brilliant Englishmen, some of whom are household names in Indian history. Among them were the Lawrence brothers, Nicholson, Edwardes and Lumsden.

The Sikhs, chafing at their defeat, joined hands with the Afghans under Dost Mohamed, the only occasion when Sikhs and Afghans were allies, and late in 1848 launched into the second Sikh war. They met with some initial success. At Chilianwala the Sikh cavalry rode into the advancing British cavalry and routed them. The battle is celebrated in Sikh annals for the 'backward charge' of the British cavalry over

40

A detail from the Ajanta frescoes. Over 2,000 years ago the Buddhists cho Ajanta as a site for a monastery, and the caves are both a shrine to Buddhi. and one of the supreme achievements of Asian art.

their own infantry and through their own artillary and wagon lines. The British soon avenged this reverse. In February 1849 Lord Gough who had been checkmated at Chilianwala scored an overwhelming victory at Gujarat. The Sikh gamble failed, and in March they surrendered their arms in a humiliating ceremony. Ranjit Singh's dire prophecy came true. *Sab lal hojayega!* (Soon it will all be red!)' he had exclaimed in disgust on being shown a map of India.

In the Mutiny of 1857–58 the Sikhs, with lingering memories of Mughal persecution, actively assisted the new conquerors against whom the Muslims and Hindus united. The Mutiny which erupted a hundred years after Plassey was more than a sepoy rebellion but less than a national uprising. Territorially it was limited largely to the north where at Delhi the phantom court of the last Mughal emperor, Bahadur Shah, still attracted the loyalty of many Muslims. In other areas such as Muslim Oudh and Hindu Jhansi the resistance developed into a war of local independence. The Begum of Oudh who ultimately fled to Nepal fought with courage and determination. So did the historic Rani of Jhansi who died riding at the head of her cavalry. There were racial excesses on both sides, and the scars of the Mutiny were never entirely healed on either side. Following the Mutiny India was transferred from the rule of the East India Company, and became a domain of the British Crown.

Thereafter followed the long line of Viceroys and Governors-General beginning with Canning in 1859 and ending with Mountbatten in 1947. Under the rule of the John Company India had also known a succession of governors-general starting with Warren Hastings in 1774 and concluding with Canning who was Governor-General at the time of the Mutiny. Three men stand out in this glittering chronicle of rulers – Wellesley, brother of the famed Duke of Wellington, who was only thirty-seven when he came as Governor-General to India in 1798; Dalhousie, the youngest of governors-general, aged thirty-five, in 1848, and Curzon who was forty when appointed Viceroy in 1898. This selective list could be challenged but to my mind these three men stand out for their initiative, their far-sightedness, volcanic energy and industry. Mountbatten, however controversial a figure, deserves a niche

41

view of Benares (now Varanasi). Probably the world's oldest living city, it is he religious capital of the Hindu faith.

in this gallery for if nothing became the British so much as the manner of their leaving, they owe it largely to this great-grandson of Queen Victoria who presided over the dissolution of the Indian Empire.

Wellesley's term (1798–1805) was notable for his consolidation of the so-called subsidiary system under which the princely states, in exchange for British military protection, were inveigled into accepting a Resident and general control over their external activities. Additionally the protecting force was paid for by cession of territory. As a British historian of the period observed, the system enabled the British to throw forward their military frontier, considerably in advance of their political one. The system was a concealed form of creeping territorial incursion.

Dalhousie (1848–1856) was in some respects an echo of Wellesley and a prototype of Curzon. He carried the subsidiary system further by blandly proclaiming and implementing his belief that 'in the exercise of a wise and sound policy the British Government is bound not to set aside or neglect such rightful opportunities of acquiring territory or revenue, as may from time to time present themselves.' As a result Dalhousie succeeded in splashing more areas of the Indian map red. Not as imperious as Curzon, he anticipated Curzon by clashing with his irascible commander-in-chief, Sir Charles Napier. The echoes of the Curzon-Kitchener clash lingered on in India many years after the two chief protagonists had departed from its shores. Napier, unlike Kitchener, was never inhibited in utterance. He dismissed Dalhousie in an angry outburst as 'weak as water and as vain as a pretty woman or an ugly man.' But Dalhousie proved more durable than Napier.

With Curzon (1899–1905) British rule in India, certainly the British administration, reached its apogee. A restless, demanding, commanding person, Curzon could not but perpetrate some blunders and mistakes. Contrary to popular belief Curzon did not initiate the partition of Bengal which is associated with his viceroyalty and name. He inherited the policy and implemented it. Nor did he inspire the controversial Calcutta Municipal Bill, which was also a legacy. However his tremendous energy and drive, despite his damaged back, found expression in some constructive if contentious reforms.

He laid down a moderate and sensible policy for the North-West frontier where he was successful in curbing the enthusiasm of the 'forward school'. On the North-East frontier he was less successful and the Younghusband expedition to Tibet in 1904 came in for wide criticism in British liberal circles as an unjustified 'little war' on a practically unarmed people. Curzon's reform of the educational system ran into a volley of Indian attack.

His dictatorial methods often embroiled him unnecessarily with Indian opinion which initially he tended to ignore. But on the credit side of his ledger are the imaginative creation of the archeological department and his reorganisation of the department of agriculture. More than any other Viceroy, Curzon made India and Indians aware of their artistic heritage. In contrast to his attitude to India's archaeological glories is the vandalism by two earlier Governors-General, the Marquis of Hastings and Lord William Bentinck. The former broke up one of the most beautiful marble baths in the Taj and had it transported to England as a gift to George IV, then Prince Regent. The rest of the marble from the suite of apartments from which the bath was taken, along with some exquisite fretwork and mosaic, was later sold by Bentinck by public auction. But for the fact that these marble fragments did not yield the money expected, it is not unlikely that the Taj would have been pulled down in the same manner and the debris put up to auction.

We are too near events to assess Lord Mountbatten's place in Indo-British history. Mountbatten has been criticised by many of his countrymen for hustling India into independence by a precipitate process of partition which engulfed both India and Pakistan in a blood bath. I cannot see what else the Viceroy could have done in the circumstances which faced him. The great Calcutta killing of August 1946 had taken toll of around five thousand lives, mostly Hindu. In the Bihar reprisal riots which followed, the victims were mainly Muslim. 'India in March 1947', as Mountbatten later declared, 'was a ship on fire in mid-ocean with ammunition in the hold.'

The pros and cons of British rule in India have still to be

carefully scrutinised. The British ruled India as a whole for 130 years. The Mughal Empire began with Babur's proclamation of himself as Emperor of Delhi in 1526, and ended nominally with Bahadur Shah's exit in 1848. By that yardstick, the Mughals ruled India for a little over 300 years. Actually the Mughal dominion in India came virtually to an end with Aurangzeb's death in 1707, i.e. within less than 200 years. Thereafter, as we have seen, the Mughal emperor survived as a shadowy symbol in Delhi. To many of India's varied races, particularly the Rajputs, Marathas and Sikhs, the Mughals were as much intruders as the British.

British rule, alongside many debit items, has a considerable number of credit items in its imperial ledger. Britain gave India a structure of stability and security, perhaps even a semblance of unity, though as political rivalries grew between the country's various communities, more especially between Hindus and Muslims, the semblance of unity was destroyed finally by the reality of partition. Perhaps Britain's greatest contribution to India was the establishment of the rule of law for it ensured the equality of all men before the law, and this in a country where previously Hindu law had differentiated between Brahmin and non-Brahmin and Islamic law observed one law for Muslims and another for *kafirs* or infidels. Britain also developed communications with the opening of roads and railways, expanded irrigation by a system of canals and waterways, and vastly improved the public health services. With the opening of communications internal industries and trades developed. The Suez canal brought Europe nearer India.

I think the most interesting phenomenon emanating from British rule was the creation of an Indian middle class. The process started with Macaulay's celebrated minute of 1835 which recommended English as the medium of instruction. Macaulay's motivation was not really as liberal as it might seem, for his intention was to utilise this new educated class of English-speaking Indians in a subordinate capacity as a link between ruler and ruled. He visualised this new class as interpreters to their countrymen of British ideas and institutions, and inevitability of the benefits of British rule. The actual outcome was ironical, for in the long run the creation of this class was detrimental to the continuation of the Raj. It opened Indian eyes to western progress and also enabled the educated Indian to rediscover India. Ultimately, knowledge of British ideas and institutions intensified the clamour for self-government. It paved the way to political freedom.

English was to help in the unifying process in India, but at first its introduction intensified the cleavage between Hindus and Muslims. English displaced Persian, the official Mughal language which was the language of educated Muslims. Hindus took to English more quickly and earlier than Muslims, who at first boycotted English education and thereby suffered in the competition for jobs. The commercial gap, as also the gap in government employment, widened between Hindus and Muslims. It was not until 1875 when Sir Syed Ahmed Khan founded the Anglo-Oriental college (now Aligarh University) that Muslims dropped their prejudice against learning English.

Thus the Hindus succeeded by their own efforts in gaining a leverage which economically they were never to lose. The headway they achieved also initially placed them at a political advantage vis-à-vis Muslims which the British tried to neutralise by various expedients, notably by the creation of separate electorates for Muslims initiated by the Morley-Minto reforms of 1909. I think this was one of the major mistakes the British made. It unintentionally and unforseeably led to the partition of India. In my view joint electorates with reservation of seats, numerically weighted in favour of the Muslims and other minorities, might have averted the final tragic outcome. Joint electorates might even, from the British point of view, have ensured an extension of British rule for inside a single composite structure, the Raj could have maintained a better and more endurable balance of power. With separate electorates, the British were merely one corner of a triangle, the other two being controlled separately and respectively by Hindus and Muslims. Separate electorates spelt a diffusion of power. Any one of the three corners of the triangle was in a position ultimately to neutralise the other two. This is what Jinnah finally succeeded in doing. He out-manoeuvred both the British and the Congress. 'Quit India,' said Gandhi to the British. What did Jinnah say? He did not plead with Britain to remain. He said, 'Divide and quit.' Against their better instinct and judgement, the British were finally compelled to do this.

For the partition of India the Muslims, in my opinion, were least to blame. They undoubtedly realised that with Hindus and Muslims moving politically in two separate, parallel lines there could be no meeting ground between the two communities. Paradoxically, as the prospects of freedom drew closer the Muslim attitude at first mellowed while the Hindu attitude hardened. Thus in 1931 at the second round-table conference in London the Muslims and Congress agreed to the principle of joint electorates but the talks broke down on the allocation of just one additional seat which the Muslims demanded. Instead Gandhi decided it should go to the Sikhs. Similarly in 1938 Congress, flushed by its victory at the polls, peremptorily turned down Jinnah's plea to form coalition governments in some provinces. With the likelihood of Britain's withdrawal from India looming ahead, the Muslims probably thought it wise to do a deal with the Hindus who in any case would be left in the majority. Jinnah's demand for Pakistan only came later and even then, until the eve of independence, he refrained from defining his claim.

Later, after 1938, Congress erred in not accepting federation. The Muslims, it is true, were opposed to the idea but the British, who were strongly inclined to it, could with the support of Congress and the princes, have induced the Muslims to toe the line. Had federation come into being it is doubtful if Pakistan would have materialised. Moreover, as part of the federal government the princes would have been in a stronger bargaining position vis-à-vis an independent Government of India and might have succeeded in salvaging their status and tenure. As late as 1946 Jinnah accepted, though with some qualifications, the Cabinet mission's proposal for a virtually federal scheme of government with a Union Centre. This suggests that his mind was even then not closed on the partition of India.

The British erred in encouraging the Muslims to organise themselves politically on the lines of separate electorates which congealed Hindu-Muslim rivalries in two separate channels and led logically to Pakistan. One cannot expect the

46

blood-stream to flow if the veins and arteries do not cooperate. In the end, separate electorates boomeranged as much on the British as on the Hindus, for partition undid the concept of a united India for which the British had laboured for over a century.

The British failed in one other respect. The Indian middle class they had brought into being finally turned against them. I think this was because until the very end they never really established a communion with the middle class whom they were prone to treat patronisingly, and with suspicion and distrust. Nehru, a product of this class, succeeded in winning the vast majority over to Gandhi's banner though many of this class, like Nehru, disagreed with the Mahatma on his economics as also on some of his political and social techniques.

In January 1948 Gandhi was assassinated. His death at that juncture really made no difference to the policies Nehru pursued after independence. Nehru and his colleagues, notably Vallabhbhai Patel, had accepted Pakistan against the Mahatma's advice and wishes, and though until his death they deferred to him and continued formally to seek his counsel, Gandhi's voice no longer exercised the old authority. On August 14, 1947, when independence was proclaimed, Gandhi left Delhi for Calcutta where he stayed with a Muslim friend. He was registering his quiet protest against partition.

In his seventeen years of power Nehru took India in many respects, particularly in the economic field, along paths almost directly contrary to those of the Mahatma. He had however, even in Gandhi's lifetime, never concealed his differences with his leader. These affected some fundamental issues relating to politics and economics and to some social attitudes and problems. Economically Gandhi had believed in working upwards from the grass roots base of rural India carrying the people with him at every stage. Nehru was more concerned with accelerating the pace of progress by imposing reform from above. He had no aversion to industrialisation along with rural development. Gandhi was opposed to heavy industrialisation principally on the ground that it threw millions of people needlessly out of employment. On that ground he was accustomed to characterise machinery as 'evil'.

Gandhi often described riches, even if enjoyed by princes, capitalists and landlords, as a trust to be promoted for the benefit of the needy and dispossessed. Nehru frankly had no patience with such esoteric economics. 'He pities the plumage but forgets the dying bird,' he once observed. At the same time Nehru was heard to comment after Gandhi's death, 'His ideas weren't silly. They brought results.' He temporarily accepted many of the Mahatma's ideas before independence as better calculated than his own to carry the Indian people together in their fight for freedom. Thus he accepted non-violence as a practical weapon in the circumstances then prevailing though he never accepted non-violence as a principle. Neither does the independent Government of India. This is often over-looked while assessing some of Nehru's actions as Prime Minister, notably his action against the Portuguese govern-ment of Goa.

Nehru was undoubtedly a Marxist though he was not a communist. He ruled out totalitarian methods to achieve what he conceived to be democratic ends. He often gave one the impression that to his mind the only difference between communism and socialism was in the methods employed to achieve the end. If the means were democratic, how did it matter? Significantly he always insisted on labelling demo-cratic socialism as India's goal, the first signifying the means, the second the end. With the overwhelming majority he com-manded in Parliament until his death, he was able to push through some near-Marxist legislation by democratic meth-ods.

Nehru's public career began in 1929 when at Gandhi's in-sistence he was chosen for the first time as President of the memorable Lahore session of Congress which passed the resolution declaring independence (as distinct from dominion status) as India's political goal. Nehru had politically arrived. He was then forty. The thirty-five years from 1929 until his death in 1964 fall into two almost equal parts. Independence came within eighteen years of the Lahore Congress. There-after for seventeen years Nehru was India's prime minister.

For a man whose favourite word was 'dynamic' Nehru's thinking on many basic problems – political, economic and social – was curiously static. His economic outlook had only acquired a defined mould after his visit to Soviet Russia in 1927 when it settled in a socialist pattern. His anti-colonialism under the influence of Marxism also acquired a strong ideological base, and his natural antipathy to capitalism was refurbished. At the Lahore Congress he declared himself a convinced socialist with no use for princes and capitalists. He proclaimed his faith in democracy and secularism, and his belief in the unity of India. His autobiography published in 1936 reiterates these views which formed the guidelines of his policies as prime minister. These guidelines were socialism, democracy, unity and secularism.

In the early years after independence the corner-stones of Nehru's foreign policy were anti-colonialism and non-alignment. It was noticeable at the Belgrade conference of non-aligned countries in 1961 that anti-colonialism had taken second place to non-alignment in his list of priorities, for by then Nehru was convinced that imperialism was on its way out. He still however saw colonialism and capitalism as two faces of the same coin, the latter leading to the former. He justified his policy of non-alignment as being the best safeguard for peace for the developing countries or 'Third Force'. As he once explained: 'The developing countries need peace for their development. They need at least two decades of uninterrupted peace.'

I think Nehru was right so long as the cold war lasted and so long as he implemented the policy of non-alignment scrupulously. With rapid nuclear development and its horrendous destructive possibilities in war a thaw set in the cold war, and from this emerged the equation of the two super-powers. With that equation non-alignment, in my view, lost most of its relevance in the world context. Today is the age of bi-alignment and multi-alignment.

In the implementation of his policy of non-alignment Nehru tended to lean heavily on one side, more often in favour of the communist world than of the democracies. This tendency placed India at a disadvantage, and over the years has lost her both friends and influence. In October 1956 when Russian tanks crushed the revolt in Hungary, India was the only non-aligned country to vote with the Soviet Union. Thereafter the credibility of India's policy of non-alignment was badly dented and shaken.

Some of its inherent contradictions and limitations were further exposed when the Sino-Indian border clash took place at the end of 1962. When the Chinese threatened to over-run the Brahmaputra valley, Nehru called for military aid from the west, from both America and Britain, who immediately responded. He also asked Russia for help. Khruschev replied in the non-aligned coin. He advised Nehru to negotiate with China without stipulating pre-conditions. At the same time he counselled Peking to show restraint. The Afro-Asian world reacted in the same measure. They advised restraint on both sides and Nkrumah, then in power, asked Harold Macmillan, then British prime minister, not to send arms to either side. The wheel of non-alignment had come full circle.

Nehru's term as prime minister could again be divided into two almost equal parts – the nine years before October 1956 when India's attitude on Hungary splintered the country's non-aligned image; and the eight years after that which saw the débâcle with China and ended with Nehru's death. Nehru never forgave China, for China had exposed him as the god who failed. The first nine years of his prime ministership were triumphant, the second eight clouded.

Yet it must not be forgotten that Nehru's radicalism, though sometimes misdirected, was always tinged with humanism. He abhorred injustice and fought inequity and reaction. He tried to liberalise the Hindu social structure and succeeded in restructuring it more in conformity with progressive life, particularly in the matter of inheritance and property for women. He could sometimes be harsh but he was never vindictive. Though he declared India's independence he was careful not to take the country out of the Commonwealth, thereby sponsoring a new concept of Commonwealth which at least initially enabled the old and new members of the club to work in comparative harmony. By never going outside the constitutional framework despite his overwhelming parliamentary majority, he inculcated a sense of democratic values in his people. Nehru believed in a free press and an independent judiciary and, though at times criticised by both, he was zealous in preserving and respecting their freedom and independence. His daughter, Indira Gandhi, has tried, as yet unsuccessfully, to encroach on both. In his lifetime Nehru managed by the charisma of his personality and the strong, cohesive government he built up at the centre to preserve the unity of India. This unity, despite his daughter's even more commanding majority in Parliament, showed signs of faltering after his death.

Lal Bahadur Shastri who succeeded Nehru died too soon. His prime ministership lasted barely twenty months. Shastri, who like Nehru, came from Uttar Pradesh, the former United Provinces, was temperamentally meek and mild, a tiny withdrawn bird-like man whose appearance was deceptive. He was by no means simple. He could be unexpectedly firm and had a mind of his own.

Shastri stood somewhere between Gandhi and Nehru. Like Gandhi he was a devout Hindu and temperamentally conservative, but he was no traditionalist and more forward-looking in his economic thinking than the Mahatma. He favoured a balance between industry and agriculture. At the same time he was suspicious of ideologies and slogan-mongering, and even during his short term gave indications of a more pragmatic approach to national planning. Shastri in his younger days had felt that Gandhi was not sufficiently militant, and politically he had always looked to Nehru as the pace-setter. He was 60 when he assumed the prime ministership.

His short tenure of office was overcast by Pakistan's two attacks on India mounted in quick succession, the first early in April, 1965, in the Rann of Kutch, and the second in August in Kashmir. Both attacks were unexpected and took India by surprise. Eighteen years earlier, shortly after Nehru became prime minister, Pakistan's attack on Kashmir had initially caught India on the wrong foot. Of the two attacks, Pakistan's probe into the Rann of Kutch, which came as a complete surprise, was the more successful. Here the Pakistani troops managed to consolidate some of the positions they had over-run and perhaps emboldened by their success decided to mount an attack on Kashmir.

Nobody in New Delhi had anticipated two successive thrusts within a matter of months. In the initial stages the Pakistani troops threatened the vital Chamb sector in Kashmir and crossed the international boundary on September 1. To draw the heat off from Chamb the Government of India decided to launch an offensive in the direction of Lahore and Sialkot, and on September 3 the Indian forces were ordered to cross the international line.

The feint succeeded. The war settled down to a stalemate during which the UN intervened. At that time the British prime minister, Harold Wilson, wrongly accused India of first breaching the international line. Wilson subsequently confessed to his error in his memoirs pleading that he had 'been taken for a ride' by the pro-Pakistan caucus in the British Foreign and Commonwealth office.

Though both sides agreed to a cease-fire a dispute arose over the return of the occupied territories. At this stage Soviet Russia intervened and invited both India and Pakistan to settle their differences round a table in Tashkent. Pakistan had overrun some Indian territory but of no strategic significance or value. India,* however, had captured two vital features in Kashmir, Tithwal and Haji Pir, which dominated Pakistan's main network of road communications. Shastri, backed by the Indian Army, was reluctant to surrender these vital points, but was persuaded by the Russians to do so. The heart attack to which he succumbed soon after signing the Tashkent agreement was induced largely by his known disquiet over his action.

Shastri's death was calamitous for the country. He was a middle-of-the-road man, influenced by no pendulum passions, and had he lived longer it is possible that the subsequent split in the Congress party which splintered that monolith might never have occured. In that case, the internal progress and development of India might have marched along smoother lines.

The result of Pakistan's second attack on Kashmir was to harden Indian opinion. Even the 'doves', who prepared to discuss the problem, now saw no reason why they should do so after Pakistan's two unsuccessful attempts to snatch Kashmir by force. They were unwilling to be coerced at the point of a gun. Indian opinion, as expressed by Parliament, has since laid an embargo on any further discussions on Kashmir which is now officially regarded by the Government of India as a settled fact, not accessible to further parleys.

Indira Gandhi's landslide victory in 1971, overtopping any which even her father had achieved at the polls, gave her the advantage of unchallenged power for five years but also a dangerous pre-eminence. The future depends a great deal on the wisdom with which she exercises the tremendous authority the country has vested in her. One disquieting symptom of her runaway victory is the decreasing public interest in Parliament where an overwhelmingly strong government is looked upon as a battering ram against a microscopic opposition. Excessive authority tends to make one authoritarian, and the prime minister's frequent injunctions to the press and the judiciary to put their respective houses in order have made many democrats in the country uneasy. Even more disquieting is the incipient fascism of some of Mrs Gandhi's over-enthusiastic followers who outside Parliament have latterly been inclined to take the law into their own hands against their political opponents.

As against these disturbing trends, there is no doubt that Mrs Gandhi symbolises in her personality the new style of the new generation of India. Nehru's death left behind him the remnants of Gandhi's Old Guard, no longer the selfless patriots of yore, but men who for seventeen years had savoured the fruits of office and relished what they had tasted. Shastri did not really belong to the Old Guard. He was fifteen years younger than Nehru and almost a decade separated him from men like Morarji Desai and Nijalingappa. When Shastri died, the then Congress President, Kamaraj, a veteran from Madras, shrewdly plumped for Indira Gandhi instead of Morarji Desai as the next Prime Minister.

I do not think Kamaraj did this with any awareness of the generation gap which Mrs Gandhi represented. He probably calculated that he could manage her more easily than he could rigid, prickly Morarji Desai. In this, he ultimately discovered, he was mistaken. The lady proved too sharp for him.

Her long association with her father as his *chatelaine* throughout his prime ministership was politically invaluable. Since the age of four Indira Gandhi was thrown into the turmoil of Indian politics, and much of her childhood and youth was lonely and disturbed, with her grandfather, father, mother and other relatives intermittently in and out of jail. She knew the country's leading Congressmen before and after independence, and with her keen observation, reinforced by listening in to some of their discussions, was probably able to assess their personalities long before she assumed the prime ministership. As her father's hostess she had also had the advantage of meeting some of the leading statesmen of Europe, the Americas, Asia and Africa when they visited India.

She succeeded Shastri as Prime Minister two months after her forty-ninth birthday. Nehru lacked Gandhi's talent in building up a second rung of leadership, nor did he designate his successor as the Mahatma had done. The first omission proved more serious than the second for the choice of Shastri as Nehru's successor to the Prime Ministership posed no formidable problems. Nehru's failure to build up a younger generation of leaders had more serious consequences. When Shastri suddenly died, the Congress was still in the hands of the tattered remnants of Gandhi's Old Guard who, because of their long years in office, were out of touch with the masses and, because of their age, were out of step with the country's youth.

Nehru's forgotten generation fell broadly into the category of those born between the two world wars. At the time of Nehru's death they belonged to the 25–50 age group. Indira Gandhi fell into this category, and as one of the forgotten generation was politically and psychologically distant from the seventy-year-olds who formed the core of the Old Guard. A clash between the two was inevitable. When it came, it led inexorably to the splitting of the Congress.

Mrs Gandhi's landslide victory was achieved primarily by her symbolic personality but also by her battle cry of *Garibi Hatao* (Remove Poverty) which attracted both the youth and the masses, but which also entailed some extravagant promises of prosperity to come. But for the eruption of the Bangladesh crisis in East Bengal and the mammoth flow of refugees into India Mrs Gandhi would by now have been confronted with her election pledges and promises.

As I write, the outcome of what the eruption in East Bengal will lead to is anybody's guess. Whatever else might be laid at

India's door she cannot be accused of starting the conflagration in East Bengal. The fuse in Dacca was lit by General Yahya Khan and his military junta in Islamabad on the fateful night of March 25, and the consequences which have followed are entirely of their making. India is not responsible if their military calculations were wide off the mark, and if the bombs they planned to detonate beneath the Aswami League have exploded in their own faces. The refugee exodus to India was a calamitous by-product of Islamabad's folly. The fount and source of the refugee influx is not to be found in Calcutta but in Dacca. The first priority therefore is a political settlement between President Bhutto's Pakistan and Sheikh Mujib Rahman and the other leaders of Bangladesh who had put their trust in the ballot and found it answered by the bullet. Peace in the subcontinent can only follow a political settlement between Islamabad and the democratically chosen leaders of Bangladesh. The people of Bangladesh seek a political, not a military settlement. One does not hunt for peace with a gun. For a non-aligned country to be aligned heavily on one side makes nonsense of non-alignment, which is why I do not agree with Mrs Gandhi's too close identification with the Kremlin. Yet I agree that Mrs Gandhi was abundantly justified in signing the recent Indo-Soviet twenty-years peace treaty after the political obtuseness of the West had driven her and India into a corner. 'I would gladly shake hands with the devil if it would save Britain,' Churchill exclaimed when Russia was driven into the war. Why shouldn't India?

A Place of Politics

James Cameron

It is worth bearing in mind, despite much that has happened since the place became a nation, and may happen yet, that India is still, so far, the biggest democracy on earth. It may be that this is stretching the definition of democracy somewhat thin, in a nation of scores of millions of illiterates, where uncounted generations have implanted an archaic and preposterous caste system that no legislation has yet succeeded in detroying, and which permanently hovers on the edge of communal strife. Nevertheless the Indian general elections are still the biggest manifestation of a free franchise in the world; no state anywhere technically disposes of so many secret votes. Every criticism of India (and much of it comes from her closest friends and lovers) has to be evaluated against that fact. While it endures it must be set on the positive side against an exasperating prospect of irrationality, venality, inadequacy, corruption, futility, and sheer well-intentioned dottiness. It says a great deal for Indian democracy that it has so far survived so much for so long. (The other paradox is that in an ancient society like India one can think of twenty-five years as a long time.) Where it has endured, even with its own curious Indian mutations, it is a monument to the domination of half a dozén people, the midwives of Independence.

This is where I come in. My erratic affair with India began in those tormenting days when imminent Independence was something to be wrenched out of history in a kind of agony of reconciliation. An Imperial epoch that had begun with blood and conquest was ending in a maelstrom of wrangling and casuistry and special pleading that I, taught in the comparative simplicities of the old India League, felt I could never understand. I am not sure that I ever did come to understand it, or indeed that anyone ever understood it. In those

56

days it seemed less important to understand all this convoluted chat than to go along with it, in sympathy and hope. The arcane argument surged around one's head like some sort of miasma charged with little darts of rancour and unexpected gestures of generosity. Only the people were real.

There is a moment, shortly after one's first landfall in India, when all one's preconceptions seem built on the most outrageous fallacies, when everything in sight undermines one's store of basic facts, when one bows meekly to the traditional cynicism of the Old Hands. Even from far away India had appeared without parallel as a source of complex and rewarding extravagances in every social, political and religious aspect. I reckoned that I had boned up on the theory of this rich mass as much as anyone could, and of course in the first day I recognised the truth of what everyone said: that India is ten million things the books fail to mention.

It was, to be sure, a time of almost unbearably mounting tension, crises, nerves and expectation. After all those years, as far as Independence was concerned, the die, in the cliché-language of the time, was cast. The principle was conceded at last, a British Labour government was in power, the Japanese war had collapsed in an abrupt and deafening silence with the shock of the atom bomb; Indian freedom was inevitable, and we were now enmeshed in a completely ferocious tangle of methodology. Freedom yes, but in the name of twenty thousand interlocking and competing deities, and almost as many politicians, *how*? Congress was overwhelmingly powerful in Hindu India, but the Muslim League was as strong among the Muslims. The 'two-nations' theory of Mr Jinnah had finally found its expression. Either could block any plan. The place abounded with strong men who, as was said at the time, could find a difficulty for every solution. Nor could the British hold the ring much longer; they had no longer the power, nor in truth the inclination, to do so. Time was running out at a rate of knots.

It is perhaps naive to assume that anywhere in the world politics is wholly a means to an end; among politicians it is bound to become an end in itself; yet nowhere as in India then did I feel politics becoming totally obsessive, governing not just what people said but how they dressed, ate, prayed,

57

village in the Thar Desert, Rajasthan.

behaved. By the time the British Cabinet Mission flew in – Sir Stafford Cripps, Lord Pethwick-Lawrence, and Mr A.V. Alexander, in curious topis of the model usually associated with Edwardian missionaries – the politicians had become almost a caste of their own, a milling contending intermingling congeries who daily grew more remote from the society they claimed to serve. This becomes a commonplace in all nations, but in India in 1946 it seemed to come about quite suddenly; one got the impression that overnight a multitude of ready-made politicians had been born – candidates, managers, fixers, entrepreneurs, PR men, all sprung by some spontaneous generation as though they had been doing it all their lives.

By now New Delhi was blisteringly hot. The permanent officials were furious; it was against all the rules that any sort of serious work should take place in the plains in high season. The city was jammed to the doors with a fluctuating tide of politicians, Princes, newspapermen, secretaries, idealists, cynics, black marketeers, beards, turbans, uniforms, sweat, Indian gin and Australian whisky and anxious fanatics of all persuasions surrounded by the thousand concentric circles of speculation and propaganda. Each day one passed through every emotion from mad enthusiasm to wild despair. It was a state of affairs I came to accept as normal – the high-keyed argument, the blend of generosity and suspicion, the over-statement of feelings that insisted on the most commonplace questions being resolved by gropings into obscure philosophy.

By and by we all packed up and went to Simla, for the final act.

Here was a new script, a new scene, a new world. There was an electric sense of urgency and climax. All of a sudden the entire company had decided that it was no longer possible to burrow its way through the entanglements of independence in the intolerable oven of New Delhi, where the heat made it hard to rationalise even such problems as whether to cross the road or not, whether to stand up or sit down. Abruptly the whole cast of this intricate road-show – ministers, advisers, consultants, Congressmen, Leaguers, Viceroy, and about a hundred newspapermen complete with camp-followers – found itself transported to this Victorian Himalayan hill station, found itself toiling up steep slopes, living in hideous proximity, gasping slightly for breath at seven thousand feet, at long last on the edge of winding up Imperial India.

This was really finding freedom the hard way. Nothing ever done in India had been so difficult. For weeks the British delegation had been talking, pleading, even threatening, trying to reconcile the irreconcilable. This was the payoff. There were no more secrets. The two great opposing camps would consider independence on no terms other than their own – Congress for a United India; the Muslim League for Pakistan: the Ulsterised India. Both attitudes had been argued endlessly, with a monstrous, monotonous insistence. Jawaharlal Nehru told me in those first days: 'After many years of experience and strife, I am almost achieving the condition of detachment.' It was one of his least convincing ironies.

In those days, and those that preceded and those that followed, there was no question but that the fate of India rested almost entirely in the hands of a group of people small enough to have been accommodated in one room – as, indeed, they frequently were. I saw them constantly – severally, collectively, at both near and far remove. Their appearance and demeanour came to have much more explicit meaning for me than that of any politicians of my own country. They were, it seemed to me, the many faces of the All-India morality-play personalised in the characters of men of such different natures that they could have been chosen by some pedantic casting director, yet involved together in a way probably no statesmen had been involved before. They existed under the unexpectedly small but predictably potent shadow of Mohandas Karamchand Gandhi.

This is no place to delve again into the confused depths of Gandhi-lore – if indeed the opportunity will ever arise again, so profoundly have the myth and mystery been explored, so thoroughly, with such enormous and repetitious detail, not least of all by Mahatma Gandhi himself, since nothing in that vast bibliography could be as comprehensive and explicit as his own *Experiment With Truth*. Gandhi the man, Gandhi the politician, Gandhi the philosopher, the patriot, the guru, the conjurer: they have all been distilled scores of time in the great adoration-machine so that what is left slips through the fingers, and sometimes I, who came near to being a cog in that

59

machine, am at a loss to remember what was reality and what was illusion.

The first time I ever saw him was long before Simla. He was in Bombay for a meeting of Congress Working Committee, and in spite of what even Bombay called a heat-wave a quarter of a million people had come to his evening prayer-meeting in Shivaji Park. Vast concourses had been common in the city then, but this to me was fantasy. When the dusty park was jammed to suffocation twenty-five thousand more rammed themselves in; when there was no more room whatever on the ground hundreds more climbed the roofs or swung from the coconut-trees. I believe to this day that it was the most densely populated section of the earth's surface I ever saw in my life. A completely solid mosaic of brown flesh and white homespun cried: '*Gandhiji ki-jai! Bharat Mathat ki-jai!*', and the little man in the crumpled *dhoti* on whom this devotion was focussed climbed on the dais among his silent familiars and apparently fell fast asleep.

As always, the purpose of the prayer-meeting was to signify the cohesion of differing faiths in a common cause: they chanted verses from the Hindu *Gita*, the Muslim *Koran*, sacred songs from the Parsi liturgy, all into a routine combining all these elements plus a touch of the Christie Minstrels. The Mahatma sat cross-legged under a bleak electric bulb like a carven figure from a shrine, breaking the illusion from time to time with sudden merry beams and bows. Finally he spoke through the crepitant microphone, in a very gentle, very old and almost inaudible murmur that had something in it both caressing and petulant. He said now was the time to seek moral strength, violence led nowhere; a string of what in any other circumstances would have been called only banalities. The huge gathering heard him in utter silence; only when Gandhi abruptly stopped and moved away did they become a seething wave of demented individuals, wheeling and swaying and chanting while Gandhi's garlanded car crept through them, horn blazing, while the sun went down behind a veil of palm-fronds, and I felt I had seen at least something of the curious forces moving towards the crisis of India.

Later he came to Delhi, in his special train of third-class

carriages only, in his customary blaze of secrecy. Once again that deprecating yet nimble figure conjured up the multitude as he always could no matter what he did nor where he might be. He had decided to live, as one of his mystic-political gestures, among the Untouchables, to symbolise his pre-occupation with the fifty-one million souls whom the Hindu system had cast out from spiritual hope: the Untouchables whom the British called the Scheduled Castes and Gandhi called the Harijans – the Elect of God. To that end he would conduct his part of the negotiations from the Sweepers' Quarters in the Reading Road. It would clearly have been too simple a proposition altogether merely to have had a sweeper move out and Gandhi move in. Much was done in the way of installing lights, fans, telephones, bathrooms, with similar amenities for his staff, making the whole set-up probably the only Untouchables' Quarter in history fitted with all modern conveniences. It gave birth to a saying that has rung down the years, and of which I – despite many rival claims – was in fact the first recorder. I visited the great man with his old friend and collaborator Mrs Sarojini Naidu, the poet, politician, socialite, ex-prisoner and wit. 'Ah,' said Mrs Naidu, surveying the elegant scene, 'if the Mahatma only knew what it costs us for him to live the simple life.'

I was to meet him again on a train journey from Delhi to Kalka, when he irascibly refused to be drawn into any conversation on matters of political moment. Instead, he would be pleased to instruct me in the correct way to peel and eat a mango without getting drowned in the juice. It was a lesson I never forgot.

It was in Simla that everybody came together, as it were in the finale of some elaborate Elizabethan drama. It was a troublesome town to work in; everyone of importance appeared to live at the remotest ends of long lanes or the topmost summits of hills; there was no way of getting about except on foot or in *rickshas* which, because of the rarity of the air and the malnutrition of the *ricksha-wallahs*, had to be pulled by four gasping men. It occurred to me to hire a very old horse, and on this I clattered about from one high headquarters to another. The only other person to whom this idea had occurred was Mr Nehru, who trotted briskly about on a humo-

61

rous piebald pony; we would bump along together talking non-committal ideology.

The magical moment of the day was dawn. I would crawl out into the pearly haze at five a.m. and ride slowly out to see Gandhi. He lived farther away than anyone, in a rather sub-urban-looking furnished bungalow called Chadwick, of all things, at the end of a five-mile track. There, as the sun touched the spine of the mountains, the High Command of All-India Congress would meet, and plan the day's undertakings, and consider, and meditate. It was a repetitively hypnotic scene. On the verandah would be Nehru, shrugging into his brown *achkhan* against the early chill; the impassive swathed figure of Vallabhbhai Patel; the mountainous bearded bulk of Ghaffar Khan, countenanced like an Assyrian chief; and curled up almost invisible among them, the little Mahatmaji. They would stand, and walk in silence up and down, com-municating it seemed through a sort of empathy, while the vapours rose from the dank lawn. It was an Old Testament spectacle. Then the spell would break, they would fold hands to each other and drift away; for a while Gandhi would sit alone until his noiseless acolytes glided around and bore him away to his baths, his food, his massage, his interminable writings.

On such an occasion I went to see him: a bleak room, pale varnished wood, bare linoleum. I had some of the solemn-sounding, meaningless questions journalists are supposed to present to famous people, a conventionalised excuse for personal contact. He answered some of them patiently, a little restlessly: 'It must surely be appreciated by now that my belief in *satyagraha*, non-violence, is an active thing, a militant thing in its fashion. It is possible for a violent man to become one day a non-violent man, but a coward, no. The political ques-tion is near to its end; let us see a little farther. All life implies some sort of violence; we must select the path involving the least. But there again it is not so simple. . . .'

After a while he slipped his feet into his sandals, extricated himself from the conversation with a brief donnish joke about the political effects of mountain air: an old man's joke. His eyes wrinkled with secret pleasure, and then abruptly lost all contact with the moment, peering about for his secretary.

Everyone seemed to come in at once; I found myself drinking orange-juice out of a teacup. I saw him once or twice more, but that was later; he was even older, and soon after that he was dead.

I suppose it is fair to say that Jawaharlal Nehru was not such a baffling character as Gandhi, though God knows he was intricate enough. Indeed he was almost certainly the most complex personality who ever rose to lead a great nation. As with Gandhi, there is the difficult problem of over-exposure; everybody has their theories of Nehru, their personal pictures of Nehru; furthermore what does one presume to say of political men who happen also to be tremendously literate, and who have written subtle, infinitely cultivated books about themselves? I came in the end to know Jawaharlal Nehru perhaps not well, but better than many; he was sometimes very kind to me and on at least one occasion, when I was returning in a state of neurotic depression from the Korean war, his personal and intellectual generosity was over and above the calls of acquaintance-ship. The Nehru family – or dynasty, as it is in danger of being called – was for some time of singular importance to me, and I do not claim to be especially objective.

Throughout all his career Jawaharlal Nehru was obviously struggling in a sort of emotional triangle. As John Gunther put it: 'He is an Indian who became a Westerner; an aristocrat who became a socialist; an individualist who became a great mass leader.' All this is true – or more accurately, alas, was. On the one hand there was Harrow and Cambridge, on the other there was the mystique of the holy cow. On the one hand there was personal wealth, on the other a society – his society – of grinding and apparently incurable poverty. On the one hand there was a crisp modernity of rationalist thought and an electric impatience with folly, on the other was a subcontinent of colossal mediaevalism and caste. Somehow Nehru, the intensely fastidious and doubt-ridden agnostic socialist progressive patriot, had to come to terms with, of all places, India. It is the measure of the greatness of his great days that he did, with honour and triumph.

There, then, in this strange *mise-en-scéne* in the Simla dawns, gathered the rest of that handful of prophets. Sardar Vallabhbhai Patel – what a name that was in those transition days. Sardar Patel was an extraordinary and somewhat terrifying man. He was then, and was to be for some time thereafter, the archetypal Congress boss-politico, the powerful fixer and organiser; he was Gandhi's liaison with the capitalist and rich big business interests that were now sponsoring Gandhi and Gandhism as earnestly as later they were to reject them. Sardar Patel could square this without difficulty with his own personal Gandhi-devotion. Once Gandhi had determined his line, however metaphysically phrased, it was Patel who drove it inexorably through committee. He was a very formidable man. The stature and importance began with his physical appearance: the massive countenance with the leaden eyes and the down-drawn lips, the mask of an Inca ruler. He would say 'I think not,' and the room froze with everyone's acknowledgment of the negative. He was total master of the political machine. His father had fought in the Sepoy mutiny against the British; it seemed a curiously remote conception. In those 'negotiating days' it was sometimes said that Vallabhbhai Patel would eventually be the Stalin to Gandhi's Lenin – with possibly Jawaharlal as the Trotsky. Patel was just about as far to the Right as he could go without turning up at the other side, and Indian politics might have been even odder than they are if he had not died when he did. After Gandhi was assassinated by a Hindu fanatic much blame fell on Sardar Patel, as Home Minister, for not effectively protecting him. This was a terrible blow to the iron man of Congress. Already he was competing with Nehru for leadership of India: the tough business-orientated right-winger against the intellectual unorthodoxy of Nehru. No one knows what might have become of India had he lived, but he did not live.

A man from a wholly different world was Maulana Abul Kalam Azad. The quiet and scholarly Azad was the most celebrated Muslim to believe in the Congress conception of a United India. He was an outstandingly learned Muslim (he had actually been *born* in Mecca) but his classicism and sense of history insisted on an undivided country. He claimed to speak no English, and all my encounters with the Maulana were through an interpreter, forever being interrupted by Azad with a sharp question over a nuance of translation.

The other great Mohammedan was Khan Abdul Ghaffar Khan, the great Pathan who seemed to me, in those Simla days, to be a sort of Moses, or Alexander, so splendid and imposing was he in his bearded enormous height; a man so vast he could afford to be quiet. He was the great soldier from the North-West Frontier who tried, hard and long and uselessly, as it turned out, to create unity on the Frontier. They called him the 'Frontier Gandhi'. All that happened to him, after Partition, was that he spent years in a Pakistan jail; since the country was to be forcibly divided Ghaffar Khan insisted on more division still and agitated for a separate Pathan state called Pakthunistan. That they reluctantly released him from prison in the end, an old and finished man, will not compensate for what both sides did to him in his later years.

On the one side were arrayed these figures, all apparently larger than life; on the other side stood one man: the implacable breakwater against a united India was the inflexible apostle of Islam in this ocean of Hindustan – Mahommed Ali Jinnah. He, like almost everyone else in this pantheon of Indian emancipation, is dead now, but he lived to see his creation come to life. It was Mr Jinnah and nobody else who invented Pakistan. Argument, pleas, threats, inducements all washed and beat around him and left him unmoved; Mr Jinnah demanded Partition and in the face of a million pressures he got it. He was a phenomenon. No one that I ever respected so much did I dislike so intensely. He existed only for the creation of an Islamic nation, and however one deplored his objective – as I felt obliged to do, since I do not like theocracies – one had to salute his metallic single-mindedness. He was a tall, thin, emaciated man with precisely articulated English and London suits so impeccable they were almost unreal; what one might have considered a fantastic dandy had one not recognised his iron will, his tactical agility, and his mastery over the millions of Indian Muslims. I spent a morning with him once; we were talking in earnest circles about an Asian future when suddenly Mr Jinnah glanced down, his expression set hard, and abruptly he excused himself. I thought I had offended him – a thing not difficult unwittingly to do – but no; shortly he returned and said: 'Forgive me; I just noticed that my foolish bearer had given me the wrong cuff links.' To anyone unfamiliar with Mr Jinnah's extraordinary pedantry this could not but have seemed the most impossible foppishness, indeed a parody; but nothing Mr Jinnah did was improvised or ill-considered, and I am still wondering what exactly it all meant. However, out of this was Pakistan invented; it was left to others to make it what we see today.

These memorable figures linger in the mind as manning that moment of history, and because of their extreme individuality they left an imprint on the emergent India to come – an imprint, some say, that endured too long, and that others still say, with fading voices, should have lasted one generation more. At any rate they, and the British of that time, found themselves presiding over a truly terrible breakdown. The Simla conference was the point of no return after which Partition became inevitable, and the mounting violence that harassed what should have been a noble transfer of power into something very near to a civil war. It will take history a long time to determine whether the British shock-tactic was right: when abruptly London announced that, in the face of these endless recriminations and insoluble deadlock, whatever happened power would be handed over – to somebody, anybody – not later than June 1948 (that is to say, just sixteen months later, come what may) and that Lord Mountbatten would be sent over to supervise the end. It was a pistol at the head of India. Mr Jinnah budged not an inch in his insistence on a Pakistan. A sort of chaos loomed. Without a settlement, government would fall into the hands of all manner of disparate parts, and India would be Balkanised – an ever-present danger, which haunts the nation even today.

So finally, reluctantly, bitterly, Congress agreed that Partition was inescapable. Finally, reluctantly, and bitterly, Jawaharlal Nehru broke with Gandhi on this point. Mahatma Gandhi, whose whole life had been spent on the achievement of India's freedom, shrugged off the accomplishment of his generation and himself, retired to the brave and solitary work of reconciling the warring peoples of Bengal, and on the glorious day of Independence Celebration in the capital in August 1947, Gandhi was not even there.

I remember it with an extraordinary intensity, because by that time I had acquired the sort of involvement with India that seems peculiar to certain Europeans, and which once acquired cannot be shaken off. Just before midnight on that steaming darkness of August 14 Jawaharlal said to the Constituent Assembly:

'Long years ago we made a tryst with destiny, and now the time comes when we shall redeem our pledge, not wholly or in full measure, but substantially. At the stroke of the midnight hour, when the world sleeps, India will awake to life and freedom. A moment comes, which comes but rarely in history, when we step out from the old to the new, when an age ends, and when the soul of a nation long suppressed finds utterance. It is fitting that at this solemn moment we take the pledge of dedication to the service of India and her people, and to the still larger cause of humanity.'

And that was that. We wept like children in the great avenues of New Delhi. For me I suppose it was *ersatz* emotion; nevertheless even for me it was the climax of something that must have been inherent in me for years: the dreary meetings in Camden Town, the street-corner work with Krishna Menon over years in North London, a multitude of petty propaganda that had led at last to the mountain meetings in Simla and finally to that unbearably poignant midnight hour.

And then, of course, the dawn came up like thunder: the Punjab burst into fire, the refugees fled from north to south and from south to north, convoys were ambushed, uncounted men, women and children were slaughtered on their way to seek sanctuary in the other dominion – uncounted indeed, but certainly around three-quarters of a million died in that terrible time because of the cut of their beard or the marks on their brow. 'I asked for peace,' murmured Gandhi at his prayer-meeting, but his words were lost in the rumble of the Air Force planes on their way to the Kashmir front.

The rest of the story is in the newspaper files, and adds nothing to my brief here. It chanced that it was India that exposed me to a curious involvement quite early in my writing life, when I was younger and more vulnerable. In a strange way India and I achieved independence more or less together.

67

To this day there are scores of Indians at whose side and in whose good company I went through those wracking days of 1946 and 1947; the bond that was built then is something I am sure can never be cut. Nor has it been, in the years between. That is perhaps the most part of this peculiar pull I have to India. If I am asked to conjure up an instant and spontaneous image of India the oddest things happen: the picture is never that of the Ajanta caves or Juhu Beach or the Rajpat in Delhi; none of the Air-India posters spring to mind except with an effort. Instead, I think of long nights spent on a small doctor's verandah outside Madras, of those hours in a railway-coach on the way to Kalka with Gandhiji, of desperate days in the divided Punjab, of lunatic students' evenings in Calcutta, of endless dialectic arguments on a moonlit beach in Kerala, of conflicts with bureaucrats everywhere that started in furious recrimination and ended in sentimental reconciliation. India presented a problem to me that few other places ever did: I could never stand outside and watch. It may not be love, but it is something very like it.

Of course I have the same conventional memories that I could ill spare, since I am a sentimentalist: the smell of dung-fire of an early evening, the contentious variety of the Chandni Chowk, the petrified melodrama of the Himalaya from Darjeeling, the aqueous eastern Venice of the lakes of Udaipur; a hundred things in the mind's eye forever. They are engraved there to compensate for the nightmare of a Calcutta *bustee* slum, the guilty pain of an enfamined Bihar, the betrayal of a corrupt or lazy official. Yet for me both are necessary for a land to have human meaning or reality; I am so nearly a romantic myself I value every experience that insists I share the wrong as well as the right. More than most Europeans I know India in its shadows as much as its brilliance; I was not asked to share the good times only.

The enormous catalogue of travel I have been allowed in my life is registered in me not in the things I have seen or the events I have had to record, but in the people I know. I have been unwarrantably lucky, in having known so many. Years ago I was luckier still; many are gone; there is no more a Jawaharlal Nehru, nor his and my beloved sister Krishna Hutheesing; no more so many people of lesser stature but equal value.

Yet so many remain – in Delhi, in Bengal, in the Punjab; above all one man in Community Services in Hyderabad, separated from me now by five thousand miles in distance and two years in time: my oldest friend in the world. If I named him I would have to name all the others. . . .

I have been very happy in India, sometimes. I have been desperately sad in India, sometimes.

You can be happy anywhere, but how can you be sad in a place you do not love?

The Raj

Paul Scott

Presumably one of the intentions of this book is to provide intending visitors to India with a survey of what India is and a body of opinion about how it comes to be like it; notes of what to look for and suggestions about how to make the most of it. My own contribution can find only a precarious place in this plan. It is as though in building a house and furnishing it ready for use someone had been asked to provide it with a ghost. Actually the ghost is already there but few visitors are psychic and it seems almost unkind to persuade them to see shadows in a country where there is so much substance.

Monuments, physical remains: these help little. India is full of them, but absorbs them. The Raj's monuments have acquired a patina of the use to which they are currently put and the characteristics of their present owners or occupiers. Even the few remaining statues of Queen Victoria have an air of belonging to an Indian pantheon of gods and demons, visible aspects of an invisible absolute. Bharat Mata, Mother India. A land of deafening noise and intense, melancholy, silence. It seems to me to have no echo.

People say that Indians are incapable of making arrangements, that they live from moment to moment and that this is the result of their fatal belief in preordination. There is something in it and it is as good a point to start from as any because the ghost of the Raj is perhaps first conjured – in its irritable or bad-tempered mood – by any visitor who is not prepared just to sit back and let things happen or not happen.

The truth is that far from being incapable of making arrangements, Indians have a genius for it; but a genius of such an impenetrable nature that it is often impossible for a westerner to understand what the arrangements are, or are for. He may know what they are for but he may not then appreciate the means being adopted to arrive at that conclusion. Or he may understand what the means are but neither know, nor be able to fathom, where they are intended to end. If he thinks he understands both what they are and what they are for, he will discover he had been under a serious misapprehension on both counts. If he has a lot of time on his hands and no objective of his own, he will be both charmed and entertained by this phenomenon. If his time is limited and his objective clear he will become impatient, then very cross, and finally hysterical. He will now get what he wants because hysteria is a form of madness and Indians revere the insane.

Another thing frequently said about the Indians is that they have almost no sense of history and no great interest in the future, and that this again is the result of a rigid caste-system and a religious attitude to the life they are presently living as being a mere stage on the road to nowhere. The British used to have a very strong sense of history, so much so that they were sometimes conscious of living it. This enabled them to deal pragmatically with everyday problems in the belief that their history would take care of the spiritual pounds if they looked after the temporal pence. There was, in fact, an interesting similarity between the Briton and the Indian and I think this is what enabled them to get on fairly well together.

If the Indian once believed that his personal destiny was a kind of mindless arrival at a state of permanent blissful oblivion and that all he had to do to get there was endure his present incarnation and act according to its precepts, then the Englishman may be said to have believed that the destiny of his country was to be great and that all he had to do, to ensure that it was, was to go on being thoroughly English and act according to the code laid down for that.

There came a day, of course, when the Englishman decided that his country could not be great while it went on controlling the destinies of foreigners and – simultaneously – a day when the Indian realised that India's destiny was also to be great and that the oblivion business was a Brahminical plot which the British in India had encouraged, much in the way they had encouraged Muslims and Hindus to feel like Muslims and Hindus and not like Indians. Consequently there was friction, not between Britain and India so much as between a new idea and an old idea of what being English and Indian meant, and since new ideas are naturally more vigorous than old ones, the old idea went to the wall and the Raj, which was now abhorrent both to modern Indians and modern English-men, was painlessly put to sleep at midnight on August 14, 1947. This is the ghost to which I am to draw attention. It is not particularly easy because like all good ones the ghost of the Raj is what is left unexplained when everything that can be explained has been.

Raj. It means rule, it means kingdom, it means the power and the glory of the ruler. To English people it means a phase in their imperial history. To an Indian farmer it used to mean a particular man, the revenue collector. Perhaps it still does but I imagine that the increased sophistication of Indian government with its departments for this and that, each with its team of investigators and trouble-shooters, has eroded the image of Collector Sahib as the sole source of providence's dispensations.

But up until 1947 anyway, if you asked the farmer what he thought of the Raj he would tell you what he thought of Collector Sahib. If you speak to an Indian farmer today and he is middle-aged he may well recall the time when Collector Sahib was an Englishman. If he is quite elderly he will almost certainly be able to do so and then, if you are English yourself, he will probably say that things were better in those days. He will say this partly to flatter you and partly because like old men everywhere he has come to the conclusion that nothing ever changes except for the worse. After such an encounter, leaving the village, you may – if you are English – feel an un-familiar glow, the effect of something subtler than flattery: the tug of an old sense of responsibility, of the good done to one's soul by doing or trying to do well for other people. The glow will not last long because the philosophy of personal responsibility no longer exerts influence on us; but while it lasts you will be in possession of a key to a mystery.

In its real sense of rule or kingdom the British Raj lasted 90 years, from 1858 to 1947. Before 1858, when India came into direct relationship to the Crown, there was a long and rather complicated process of mercantile escalation, begin-ning with the formation of the East India Company in 1599.

72

Top: *The gateway to Fatehpur Sikri, built by the Mughal Emperor Akbar as his capital in the 16th century.* Bottom: *The Basilica of Bom Jesus in the 16th-17th century Goan capital of Velha Goa. It houses the body of St. Francis Xavier.*

Quite briefly what happened was what you may see happening any day in industry and commerce: your small competitor securing a piece of a potentially valuable market and building from there using fair means and foul, motivated not by personal greed so much as by the logic of his activity which is that if he has a piece of the market there is no actual reason why he shouldn't have the whole of it.

It has been said that the Indian empire was acquired in a fit of absence of mind; but it was a prolonged fit and, if encouraged by the more easily arrested absence of mind of weaker competitors (the Portuguese, the Dutch, the Spanish and the French) and of the fading power of the Mughals who ruled India when the British first went out there, it was kept going most by recognition of the economic law that the best way to market goods is to own the raw materials and the means of production; in this case the land itself. Unfortunately, to own land you have to govern it, which is a full-time job. In 1833 the East India Company was quite correctly deemed incapable of trading and governing and its trading charter was taken away.

While trading passed into the hands of uncoordinated private enterprise, Parkinson's Law immediately came into operation in regard to the old company. Told that its sole concern was the government of the territories it had acquired, it looked round for more to acquire so that it should be fully occupied. The period between 1833 and 1857 was one of such single-minded imperial expansion that the Indians took fright at last and rebelled. In 1858, after the rebellion had been put down, the situation that existed was formalised. The government of British India passed from company to Crown, which also declared itself paramount over any Indian princes who still ruled their own states. The company became the Indian Civil Service and the Raj was born. It was an imperial takeover. The British electorate became shareholders and had representatives on the board (Parliament).

The analogy is of peculiar interest. It is commonly known that shareholders are characterised by a total indifference to everything that goes on in their company so long as they get good dividends and are regularly supplied with glowing and optimistic reports which bear some relationship to the state

73

view of Dal Lake, Srinagar, in Kashmir.

of affairs revealed by their half-yearly cheques. One of the most significant factors in the history of the Raj is this indifference of the British electorate; an indifference every so often shaken into surmise by news and evidence of things having been made a mess of (stormy extraordinary general meetings) but by and large prevailing from the beginning to the end. But the more immediate point of interest in the analogy is the light it sheds on the gulf that grew between the people at home (the shareholders) and the people on the spot (the officers of the company).

I think it is the Americans who first formulated the idea of there being such a person as 'a good company man'; a man whose attitudes are moulded by his duty to the company into naivety of such an intense and sublime kind that it is elevated above loyalty to the same sacrificial level at which patriotism exists. 'Don't knock the product,' the good company man will tell his wife; and mean it. The same man, failing to sell the product to a potential customer, observes that he has not failed in his duty to sell but in his duty to provide the customer with what the customer both needed and wanted.

Psychologically this is not as silly as it sounds. The cynical shareholder may sneer at the product all the way to the bank; but sourness and satire are no help to the sales director earning a living. Above all we like to feel that what we do is worthwhile. Perhaps when it is not it is better to pretend it is. Artificially inspired belief that the product is great is probably better in the short run for the human spirit than the possible alternative: the belief that nothing is great, that everyone is in everything for what he can get out of it.

There are a number of reasons why for so long I have been interested in the Raj; but one of them is my impression that in a very specialised form the Raj was representative of what the English people were when they last believed in the value of their product and in their own peculiar value; and that the act of putting the Raj to sleep was no less indicative of that belief than the longer act of keeping it alive. For me, thoughts about the Raj, the end of the Raj and the subsequent English experience of a world in which the Raj no longer exists, combine to form a moral dialogue such as any Englishman might have with himself anywhere. To that extent the ghost of the

74

Raj haunts Britain as well as the subcontinent; although the latter is the place where we more clearly came to the end of ourselves as we were, because there we caricatured ourselves in cartoon images of the Sahib and his Memsahib, twin figures of fun, bearing the regalia: the topee and the fly-whisk. These images, seen to be redundant before they were, duly became so. The Sahibs and the Memsahibs are dead and gone. But these are not the ghosts who haunt the corridors of our minds and of the Indians' minds. The true ghosts are the ghosts of the ideas that created the figures and then destroyed them.

We can fix the date the Raj began and the date it ended and say what it was and how it performed at any one period. We can plot its peaks of power, its ascent and its decline and in the process expose a number of ideas that seem to lie explanatorily at the centre: for instance that Britain had a position to keep up as the world's richest trading nation, that she had a civilising mission, a policing duty, a paternal obligation to less fortunate nations. Turn and turn about India might then look like a source of riches to be exploited, a land of heathens to be led to the light, a country of unbridled criminal passion to be subdued and corrected, a bastion to be held against the yellow peril or the Russian bear; or the home of simple uncultured peasants who needed help and guidance.

But common to all these ideas was the one that the English had of themselves as being capable of pursuing them, of being experts in every practical matter under the sun: commerce, decent living, law and order, power and politics, to name but a few. It was the result of an explosion of national pride such as I suppose hadn't occurred since the time of the Tudors.

Of course no Englishman would say, 'We are great, we know everything, we do everything that much better than anyone else.' Greatness was a mystical, inward, knowledge and best supported by a deprecating outward modesty, an appearance of ease that amounted to languor, of being awfully ordinary, as greatness was when it was shared by a whole nation.

Only abroad could an Englishman allow some consciousness of his superiority to show, and then showing it was a duty. Abroad the Englishman was an emissary, charged with his country's trust but exposed to un-English influence and under a personal obligation to hint at the consequences of any insult to his nation through an act of incivility to his person.

There was, additionally, the effect a place like India had on a man already tense with this consciousness of his private and public position and of his duty to maintain both with dignity. Leaving aside entirely the element of corruption that is supposed to enter a man's personality when power settles on him (and an Englishman coming to India automatically acquired power as a member of the ruling race), I cannot doubt that the sight, the smell, the experience of India, of Indian manners, attitudes, heightened the natural feeling an Englishman had of being English to a point where it stopped being a feeling and became a physical and mental reassurance, a ready-made suit of armour. Nowadays we call the adverse effect of encountering an alien environment 'cultural shock'. Originally we described it as 'feeling a bit homesick', which meant feeling a bit put off, even a bit frightened, and led a man reasonably enough to seek the company of his fellow countrymen in the sanctuaries that had sprung up to provide him with comfort and relief and the essential conditions in which being English could be seen to be a real state of affairs and not an illusion. Exactly the same thing happens today, for example on the Costa Brava.

From these sanctuaries, constant sources of refreshment (the bungalows, the cantonments, the clubs), one could sally forth and deal with the alien world one was required to deal with; and these dealings were also heighteners of the consciousness, the necessity, of being English; for here were corruption, bribery, false witness, a wild and irritating inability to do even the simplest thing right first time (i.e. the English way), endless argument, open emotionalism, revolting personal habits, noise, squalor, filth, religious bigotry, idol worship, ghastly practices, fawning, flattery, terrible cheek and sullen insolence. What could be worse?

Quickly, upon the pink, relaxed and jolly English face another was superimposed like a mask. Sahib's face. This was neither suntanned nor (as in the cartoon) reddened by drink, spiced food and self-indulgence. It was pale, grave, withdrawn; and the mouth when it smiled did so with reluctance so that the corners, anchored by the determination not to see anything funny in what was patently very unfunny, turned down instead of up. The movements of the limbs became slow, studied. Heat and humidity contributed to the lethargy but it was mainly a reflection of the deliberate slowing down of responses as an understanding was reached of the very narrow channel there was through which one could steer a safe course between the rock of contagion and the more dangerous rock of irreversible error of judgment and action.

There is always a difference between the man who sits at home and the man out there on the spot who represents him. They begin by seeing things in the same way but this doesn't last. The man at home continues to enjoy a narrow view on to a large prospect and a grand objective. The view of the man on the spot broadens as his eyes are opened to all the obstacles and realities that make the task more difficult than he ever imagined so that in order to perform at all he tends to confine himself to the job he has to do and leave the grand objective to look after itself.

This is what happened to the Raj. But something else happened too. Its members became good company men. Faced with what looked like incontrovertible evidence of the need for the product they found themselves believing in it with an almost passionate fervour. And they went on believing in it in its original form through all the changes made by the top management to give it a new and improved look, and were still believing in it when the management decided there was no market for it and went into voluntary liquidation.

The product of the Raj was, of course, rule; rule in the form of benevolent despotism or, as it was called, paternalism, which meant that it was supposed to be stern but just. It was a Victorian concept. In India it endured into the 1940s. In England belief in it did not survive the Great War but it lasted until that war began which is why for me the Raj does not convey a musty Victorian interior but a perpetual Edwardian sunlight.

When we speak of paternalism in connexion with the Raj we misjudge its nature if we think of a bearded patriarchal figure, bible in one hand and sword in another, ruling the natives with both. There was, before the Mutiny, paternalism of that kind. Indeed with its hints of forcible conversion to the English religion it contributed to the unrest that came to a head with that rebellion. But afterwards India became very much a young man's country.

Because the English people had no doubts about their superior capabilities, they saw nothing extraordinary in an Englishman in his early twenties holding a position of power and authority over the inhabitants of several hundred square miles of foreign territory and subsequently greater authority over several thousand square miles, for example an area the size of Yorkshire.

Nowadays we may marvel not that it was often done well but that it was done at all. Forty years ago, that young man next door who has just come down from Essex or Warwick and is going into ICI or advertising to sell soapflakes, might have been on his way on the P & O to what, it must be admitted, was potentially a considerably larger experience of life because what he faced was direct and personal responsibility for the welfare of thousands of people. That it was wrong I have no doubt. Morally it was none of his business, and I fancy he was not as well equipped as he thought; but paradoxically the moral challenge it offered was of a kind and a scope nowadays without parallel. There are greater and lesser moral challenges but none precisely like this. You have to sit and think carefully, not how the job was done, or why, or whether it was wrong or right, but of the fact that it was there to be done; and then the ghost of the Raj stirs, as if summoned, and turns its pale face to you and seems about to speak, but doesn't.

There are so many layers of prejudice, bad manners, vulgarity, pretension, snobbery, grandiloquence, sheer bloody-mindedness and cold reactionary cruelty for it to speak to you through and hope to be understood. The poor ghost has always had a bad press. And when it wrote about itself all it seemed to have to say was that the duck-shooting was good and the servants were faithful.

British India was divided into provinces, the provinces into divisions, the divisions into districts and the districts into sub-divisions. At the top of the scale was the viceroy who was also Crown representative to the Indian princes. There were over 500 of the latter. By and large they ruled their own considerable or miniature kingdoms (about one third of the total landmass) under the eye of the Political Department – an élite body that drew its members from the ICS and the Indian Army (and in the end went up almost literally in smoke, thoughtfully burning the thousands of confidential files it had kept on the sometimes erratic behaviour of its royal charges).

The Raj provided a comprehensive service. From the ICS came its civil administrators, magistrates and judges. The Indian Army offered defence and occasionally imposed martial law. The Indian Police kept civil order. A member of any one of these branches was demonstrably a member of the Raj. There were other arms in which a man could serve (e.g. medical and engineering) but the ICS, the Police and the Army formed the triumvirate upon which the whole system depended. The evolvement of limited forms of parliamentary democracy never altered this basic structure.

The unit of administration was the District (I gather it still is). There were 250 of these in British India. The average size was 4,430 square miles. Many were bigger. 'Mymensingh holds more human souls than Switzerland. Vizagapatam both in area and population exceeds Denmark. In the United Provinces each Collector is on the average in charge of an area as large as Norfolk and of a population as large as that of New Zealand.'* To rule any one of these our fresh-faced young Englishman might aspire.

The executive head of the District was the District Officer (ICS) known as the Collector or Deputy Commissioner or District Magistrate. 'He is for that area, the principal executive officer of Government and the chief magistrate. His duties are many and varied and he has wide statutory powers. He is responsible for the good government of his charge and for the maintenance within it of peace and order.'

The judicial head of the District was the District and Sessions Judge (ICS) 'whose duties extend over both the criminal and civil courts in the District, (he) exercises the highest original jurisdiction in the District and also appellate jurisdiction over both magistrates and civil judges.'

* From the Montagu-Chelmsford Report.

The third man with continuing district responsibility was the District Superintendent of Police (Indian Police) who was 'the right-hand man of the District Officer in all that concerns the maintenance of public tranquillity and the control of crime. He is at the same time executive head of the District Police Force.'

The Army, which otherwise led a peacetime life of its own, a boring one by comparison, alleviated mainly by social and sporting activities, came into the administrative picture when the Civil (as it was collectively called) came to the conclusion that it had lost control and applied for military aid. In extreme cases of civil breakdown the senior military officer in the area could proclaim martial law.

In the Raj's social scale the Indian policeman was less *pukka* (proper, superior) than the Army Officer and the Army Officer (he would not have agreed) less *pukka* than the ICS man. The typical candidate for the ICS before the last war was a young man of good family, with a university degree, the ICS examination was competitive (and eventually opened to Indians). The minimum and maximum ages for candidature varied from time to time. After acceptance into the service a probationary period had to be spent in England (a special consideration, in the case of successful Indian candidates who sat the exam in India after that, was allowed). This period also varied, so that with the variation in age limits and probationary period the ages at which a civil servant *arrived in India* 'ranged between minima of 18 to 23 and maxima of 21 to 25.'*

During the probationary period at home a study was made of Indian legal codes and acts, Indian history, a vernacular language, riding and hygiene, plus a classical Oriental language or Hindu and Mohammedan Law. At the end of his probation the candidate had to pass another exam. Arrived in India and posted to his province he was likely to feel that nothing learned so far was the slightest use to him.

Taken under the wing of the District Officer he saw and heard something of what it was all about. At the end of six months he faced a departmental examination and probably didn't pass first time. Once he had passed he became a magistrate of the second class. Six months later he sat another

* The ICS by Sir Edward Blunt. Faber, 1937.

'departmental', success in which made him a first-class magistrate and sub-divisional officer. He had his first personal area of jurisdiction. He was on the way to the Olympian heights. In due course he would have to decide whether to stay in the executive or transfer to the judicial branch. In the executive he could become a Collector, perhaps a member of the Viceroy's executive council, even Governor of a Province; or, if he preferred his court work, a judge of the High Court. In any case, as may be gathered, he was unlikely to be a fool educated only in the holidays from Eton.

The descriptions of the status and duties of Collector, Judge and Police superintendent which I have quoted do not come from the yellowing pages of an old Sahib's cosy reminiscences, but from a pamphlet published by the British Government in 1945. The pamphlet was issued to attract suitable men (English and Indian), then in the services, into the ICS, the Political Service and the Police. The only suggestion that things might not be as attractive as they were made to sound was a remark in the foreword by the Secretary of State to the effect that in view of HMG's declared policy of promoting full self-government in India and Burma, constitutional developments might necessitate appointments under the Secretary of State being terminated. 'At the same time it is my belief that when that time comes opportunities for continuing to render good service to India and Burma will not be lacking to men whose devotion is to the peoples of those countries and *whose capacity for efficient and disinterested service has already been proved.*' (My italics.)

How long would it take to prove such capacity? Clearly the Raj imagined either that a young Englishman recruited in 1945 would prove it extraordinarily quickly or that self-government still lay some comfortable number of years ahead. This pamphlet has always struck me as an embodiment of the Raj's illusion of its own permanence.

At the back of this illusion lay the old belief in the value of the product and India's desperate need of it. Soft or hard sell, not to sell was to deprive India of what she demanded whether she knew it or not. Through all the stages of constitutional development, reform and 'Indianisation' of the services (which meant admitting suitable Indians into the seats of mystery and power, the ICS, the Army, the Police), the illusion of permanence was fed because what Indianisation really meant to members of the Raj was the Anglicisation of Indians, their metamorphosis into what Macaulay called brown Englishmen. When the only difference between Harry Smith and Hari Singh was the colour of their skin the Englishman could then, yes, with a sigh of satisfaction, withdraw. What would remain forever enshrined was the product: the skill, the attitude, the cool judgment, the emotional detachment, the greatness, the whole mystique of being English and making things work as only Englishmen could.

Hard and soft sell were, I think, more successful than the average member of the Raj was prepared to admit at the time. You have only to enter an Indian Army mess today or talk to an Indian Inspector-General of Police, or to an elderly Indian still proudly insisting on the letters ICS after his name to distinguish him from the lesser Free India breed of IAS, to recognize the truth of Mr Muggeridge's remark that the only Englishmen left in the world today are Indians. The ghost of the Raj is manifest in them too. But it is no more than a ghost. They do not enjoy the same privileges because they do not have, in a parliamentary democracy, the same power.

When it came to privilege, any Englishman in India was, *prima facie*, a member of the Raj. It was from these privileges, the abuse of them, that the Raj got its bad name. If the privileges had been all that mattered the bad name would have been thoroughly deserved and would now be the only one that is remembered. But they were not all that mattered and many Indians today will tell you so.

With the privileges went responsibility. The peculiar quality of privilege is that it can be seen to be enjoyed and seen to be abused. Almost any middle-aged or elderly Indian will have a tale to tell of being cold-shouldered, insulted, thrown out of a first-class compartment on the Frontier Mail, kept waiting on a verandah, pushed out of the way in a post-office, refused admission to a club. Responsibility, on the other hand, is not something that can be seen. It is felt by the man who has it and perhaps eventually recognised by the men in whose interests he has accepted it. It is fairly unlikely that the Indian who still smarts from an old injury to his self-esteem will have another kind of tale to tell, of the Deputy Commissioner having taken a bribe, of his assistant having found for a claimant because he coveted the claimant's daughter, of a civil surgeon selling the drugs in his hospital for private gain, of the District and Sessions Judge giving light sentences to Muslims because he disliked Hindus, of a member of the provincial secretariat steering a contract through for a consideration. He will more likely tell tales of presents returned with thanks, of the unutterable boredom of pleading a simple case in front of a young English magistrate who wrote down every word in a painful effort to get both sides of the question clear, of cold refusals to discuss, visit or in any way descend from the lofty eminence from which it was considered proper to deal, scrupulously, with every matter.

There was a word for it: integrity. If it often protected itself behind implacable barriers of privileged seclusion, that did not alter its nature, nor its effect on the administration. It meant more than mere honesty. It meant the incorruptibility of a man's whole being in relation to the job he was doing. It became a passion, an obsession. It was possibly the one part of the product that stood the test of time, of history, of affairs which on other levels could have been better, more intelligently, arranged to convey human warmth. But integrity is a word that has always had icicles hanging from it. In the Raj's day it was the password to a club within a club, one to which Indians were admitted: a chilly hall, entirely devoted to business, where your Indian Member of Council, your Indian Minister, would find no cause for complaint in the behaviour of the English departmental secretary, nor your Indian District Judge feel himself belittled by the Deputy Commissioner. But at the door into the street English and Indian parted and in the street your eminent Indian Member for Education might find himself jostled by a callow English subaltern, a perky British sergeant, or his taxi taken from under his nose by an English lady suffering from myopia. In the street, as it were, privilege became the prerogative of the British Raj and responsibility a question of skin pigment.

Nowadays there is a commonly held belief that most of the Raj's faults arose from the attitude of its women. Certainly no one was more adept at making an Indian feel like something crawling from under a stone than Memsahib and she it was who, creating a home from home and solemnly observing the rites of Camberley in the heart of Mudpore, ensured that the community of the Raj should never develop along lines that could lead to an English sense of identity with India. But the Raj did not come to an end because it treated Indians badly, and having given the subject quite a lot of thought I've come to the conclusion that the women made a vital contribution to the important and lasting impression India has of the way an Englishman did his job, because of the indispensable part that used to be played by women in the Englishman's image of himself.

Do you remember? She and she alone really brought out what was good in a man. Sharp-tongued, hard-bitten, handy with rifle and gun she might be – he might hate the one he had – but the notion that a woman was 'sort of sacred' was inseparable from the image, the package, the product. Without her I doubt he would have sold it as well as he did.

How old-fashioned it sounds. But then so does the whole product, now. It wasn't old-fashioned in 1858 and when you think of the temper of England in the summer of 1914 you realise it wasn't old-fashioned then, either. The shareholders still knew little about India but were sure the people out there were doing well, teaching the Indians how to do things the English way and think English thoughts, like that clever Indian lawyer, Mr Gandhi, who went round London in the early months of the war with Germany persuading other clever young Indians to join up and fight the Hun.

If you stopped the clock at this point you would find overall a condition of reasonable harmony between the Raj and the English at home and perhaps just one significant area of disagreement. Whereas the English at home assumed the Raj was doing its job and turning out clever Anglicised Indians (the only kind they ever met) who could one day take over, the Raj took a different view, one that inclined them to believe that too many of the clever Anglicised Indians were exactly what India did not need because they were being Anglicised in the wrong way, being fed with ideas that questioned the value of the product and which were threatening to influence young people at home, too.

You might suppose that as individual members of the new young English community came out to become members of the Raj they would have brought their new ideas with them. They did. But they did not survive. The rules of the club were too strict and after a while the newcomers saw the reason for this. India was not composed of clever English-speaking chaps who played cricket, read Shakespeare and studied law. That pleasant fellow one had met at Balliol seemed different when he turned up in Mudpore, claiming acquaintance. It was pretty rotten not being able to treat him as decently as one wished but one had to be careful. He might want something it would be impossible to give. For the job's sake you simply had to be detached, disinterested and think of the country as a whole. The Indian chap's parents were stinking rich, too, and probably didn't care tuppence for the poor old peasant who called you Sahib. And then of course Hari was a Hindu (a Rajput at that) and most of the peasants in this sub-division were Muslims.

Central to the Raj's image of itself was the belief that it held the balance of power between otherwise irreconcilable forces who would lose no opportunity to cheat, threaten or slaughter one another. The political impasse between Muslims and Hindus during the negotiations for British withdrawal and the bloody events that accompanied the birth of the two new independent dominions of India and Pakistan, seemed to prove that the Raj had been right all along. But against what looks like evidence of sound judgment in this matter one has to set the fact that no attempt was ever made to encourage the two communities to work together. Instead there was a long and not at all creditable political history of playing one off against the other. The Raj always said it had united India. It had in the sense that it imposed a single rule of law upon all its people but a united India in the emotional sense was just what the Raj feared most, whatever rosy picture it had of the distant future it worked at snail's pace to secure for its charge; although to give it its due its fears were based, in this century certainly, on the grave suspicion that the Indians were still incapable of correct (i.e. English-style) government. It therefore seemed best to allow the divisions, and rule what was divided for as long as that seemed necessary.

Gandhi understood this situation very well, which was why he tried to promote in the Indian mind an ideal of Indian nationality, an ideal of patriotic service and an ideal of non-violence. If he sometimes despaired of making an Indian feel an Indian first and a Hindu or Muslim or Sikh second, he also feared the consequences of succeeding. He knew the violent nature of the people of the subcontinent. So did the Raj. The non-violent image might take in the people of England but it never took in Collector Sahib.

After the 1914-18 war the divergence of opinion between the people at home and the members of the Raj about what Indians were capable of became so wide that the two lines of belief never met again. At home the idea of what an Englishman was had begun to undergo radical changes. There in your Bloomsbury bed-sitter you sat round the popping gas-fire toasting crumpets and observed that the old world was dead and the old England with it and that it was quite wrong to go on exploiting countries that were capable of exploiting themselves. But in Collector Sahib's bungalow things looked very different, because Collector Sahib sat on his verandah and gazed at his darkening garden, sniffed the ancient smell of cow-dung fires in the ancient villages, thought of his 4,430 square miles of territory and the souls entrusted to him and observed that *his* world hadn't changed at all. Whether he thought this was the Raj's fault or India's fault was hardly relevant to the situation as it was, as he saw it. Besides which, of course, a job is the most difficult thing in the world to give up if it is one in which you believe.

About the Raj now there descended a certain melancholy, the melancholy of a people who had to face a stiffening opposition in the country of their voluntary exile and the more disagreeable opposition which came from home. Feeling itself unloved, the declining years of the Raj were, I believe, years in which the remote affection always implicit in paternalism became a different, deeper, emotion. Too late. In any case the Raj went on being correct, English, inarticulate. It expressed its love in the way it expressed everything: plodding on without change of manner or expression, incapable of the gesture that would enlarge and release. The politics of the period reveal nothing of this. There, in Delhi and Simla, were

86

the pomp, the power, the continuing process of resistance to any change that would erode the area of the influence that existed and therefore had the dynamic will to survive, which all human mechanisms have. But we are after the mystery, not the mechanism, and that in the end came down to the idea a man had of himself and of what he was in the world to do.

Sahib and Memsahib loved India more than did the man and woman back home whose parliamentary vote in 1945 ensured that the Indians would take it back. But then love has always had a possessive element in it. Dispossession now lay at the heart of the English notion of what it was to be English. In 1945 the English went in a big way into the moral leadership business. This collapsed rather dismally round about 1950 when it was discovered that moral leadership without the power, means and will to impose it on people who needed it was a bit of a pig in a poke and that in any case to lead anyone, even yourself, you have to know where you are going and why; and about this the English were suddenly unsure and still are. Our whole vocabulary and the ideas that used to give words their credibility are undergoing a process of re-examination. The word authority, for instance. What does it mean? What should it mean? Can it ever be anything but an excuse for one man telling another what to do?

India gained independence because the belief that the exercise of English authority over her was both wrong and unnecessary had become, between the two world wars, one of many beliefs the restless English at home came to have about themselves, about the kind of people they really were and about what they were in the world to do and not to do. Any tendency they had to overlook India's future when planning their own was corrected by constant reminders given by the stubborn nationalist campaign in India itself. In a parliamentary democracy, of course, no belief can become a reality until it had become acceptable to a majority of the electorate as part of a package deal for a new and interesting future. The Labour landslide of 1945 was an expression of the hope for just such a deal. But out of each 1,000 voters perhaps one was pleased that the Welfare State at home meant home-rule for India as well. The other 999 couldn't, as the saying then went, have cared less. But they had voted for it. It was part of their new government's mandate. The fascinating thing about the demission of power in India was the note of hysteria that entered the negotiations when it began to look as if the Indians would never agree about how to take their country back.

Finally, with one swift stroke the desired disconnection was achieved. What the English electorate had done was to float India off (in two parts) into seas they themselves were riding with confidence – the stormy seas of the mid-Twentieth Century. Under the Raj, India had been anchored in the backwaters of a vanished era. Its members now stood beached, surrounded by the detritus of their cancelled occupation and in possession of nothing but their unchanged belief in the value of the product which they had watched devalued under their very eyes: new and improved beyond recognition – India divided into two and already red with the blood of communal massacre, and the century-old treaties with the Princes thrown away like mere bits of paper. That the major promise, independence, had been fulfilled was cold comfort. They wished their lost charge well, but had forebodings. Perhaps many of those who are still alive think they were right, that things are not ordered as well as they used to be. But the world for which the Raj prepared and conditioned India is not the one in which, like the English, she struggles to survive.

The ghosts you will meet there are not ostentatious figures. The purple, the crimson and the gold were mere trappings. They come in the shape of a man and a woman in early middle age, he in a plain chalk-striped suit with wide lapels and crumpled trousers, she in a cotton dress. They look so ordinary that you wonder whether they ever did any of the creditable things they claimed or any of the discreditable deeds of which they are accused. You may even wonder whether the integrity wasn't an illusion too. But in the end integrity seems to fit because nowadays that is what we say people have when they don't seem to have enough of anything else to get by, when they do good work withdrawn by circumstances or temperament from contact with those abrasive elements that make life an emotionally and intellectually dynamic, dangerous and altogether very uncertain affair.

88

The tiger, perhaps India's most spectacular wild animal, is now the subject of *great international conservation campaign.*

Princely India

Manohar Malgonkar

When, in the autumn of the year 1858, Queen Victoria announced that the British Crown had taken over the running of the Indian Government from the hands of the East India Company, India ceased to be a trading company's private business and became 'The Raj'. The Royal proclamation halted a process of conquests that had gone on without interruption for precisely a hundred years since Clive had won the battle of Plassey and which, in large measure, had been the cause of the Great Revolt (or 'Mutiny' as the British rulers called it) of 1857. Halted it, as it were, in mid-stride, while there was still much left to conquer: fully two-fifths of the subcontinent still remained in the hands of what were called 'Native Princes'.

Among these princes, in addition to the fifty or so Mughal or Maratha vassals who had ruled their domains for decades if not centuries, were petty war-lords, bandit chiefs, village strong-arm men and other fly-by-night adventurers who had helped themselves to bits and pieces of the crumbling Mughal Empire. The Queen now gave an assurance to all these rulers that she would 'sanction no encroachment' on their domains, and that she would respect their rights, privileges and honour 'as our own'.

Thus by Royal command the confusion created by a national uprising and by the severity exerted in its suppression, was quick-frozen into a permanent settlement. Within an India governed by the British, another India was marked out and fenced off, and in the process, an altogether new class of rulers created.

In point of fact, these 'Native Princes' would have had little hope of surviving long, for the East India Company's hunger for land was known to be insatiable. Whether you were friend-

89

*e High Court building in Chandigarh, the new capital of the Punjab built under
* leadership of the late world-famous architect Le Corbusier.*

ly or hostile, sooner or later the Company was bound to seize your domain on some pretext or other. Had not Lord Dalhousie 'annexed' the altogether friendly and docile kingdom of Oudh because he believed that he would be guilty in the eyes of God if he did not do so? The revolt could only have whetted the Company's appetite.

These rulers, waiting to be demolished in one clean sweep, were now told that their domains were, quite literally, indestructible and irreducible no matter how they were administered. The worst that could happen to any one of them was to be ordered to abdicate in favour of his successor which, in any event, ensured that his heirs would continue on the throne in perpetuity.

Technically they did not come under the Indian Government at all; nor did they come under the British Government, for Parliament had no power to enact laws that would affect them or their subjects. They were subordinate only to the Sovereign, and to the Viceroy, not because he happened to be the head of the Indian Government, but in his capacity as the representative of the Crown. The Viceroy, on his part, exercised this tenuous control through a particularly glamourous hierarchy of officials known as the Indian Political Service. The officers of this service were chosen ambivalently from both the Army and the Civil Service, as though for the membership of some exclusive club, by a mysterious system of grading which took into account their breeding, charisma, coolness under stress and proficiency in sport as well as scholarship and administrative ability. Needless to say, they were also the highest-paid officials in India.

Even though subordinate to the British Crown and its officials, the Indian princes were, in many respects, far better off than most independent kings ever were, and certainly they were far more secure. Seldom was authority more absolute, or the divine right of a ruler more unassailable. As Mr Gandhi observed, a prince could commit murder and still 'not come under any law'. If his subjects rebelled, British troops would march in, fire a few shots to scare the rebels, and restore his authority, leaving him to torture or kill the rebel leaders at leisure. Misrule in itself was not sufficient cause for Viceregal intervention, and indeed there was nothing in the

90

treaty or covenant of a prince which required him to rule well. In a land of sacred cows, the British Sovereign had created her own, to be fattened and pampered and protected from harm even if not to be actually venerated.

The older British officials, those who had served the Company's interests well, were scandalised. To them the prohibition against further territorial gains was like blocking the current of history itself. At a different, more introspective level, it tarnished the image of the Empire-builder by calling into question the very basis of his presence in the country: that he was here on a mission of deliverance, to bring to the oppressed, abused, demoralised population the benefits of civilised government. Now civilisation was to be withheld from at least a third of the Indian population. They feared dire consequences. The only useful purpose the princely states might serve, Lord Elphinstone predicted, was to act as 'sinks to receive all the corrupt matter that abounds in India'.

The Indian rulers, on their part, did little to disprove such unsavoury forebodings. Sealed off from the outside world by the dictates of Imperial policy, they rapidly slid into an Arabian Nights kind of feudalism and surrounded themselves with all the trappings of mediaeval monarchs. Jesters, astrologers, miracle-workers, pimps, perverts and eunuchs gravitated to their courts and found employment and even positions of honour. The eunuchs had a practical function as guards for the well-stocked harems which were regarded as a dual-purpose hallmark, testifying to a ruler's potency as well as to his wealth: he had to be something of a sexual athlete to cope adequately with scores of amorous women, and really wealthy to afford them at all.

The winds of change did not penetrate their realms and still less their palaces, and power without responsibility bred indolence and fattened egos to grotesque proportions. The present held no challenge, the future no threat; so they tended to hover close around the time of their creation, the midnineteenth century, when kingship and divinity were vaguely analogous. In any case, many of them had all along laid claim to celestial origin, from the sun or the moon or some other refulgent planetary body such as Mercury or Venus, or from Agni, the god of fire, or even as the result of a cataclysmic

91

mating between the god of fire and a range of mountains. Genealogies were pruned and straightened and polished, and purity of line became a fetish, something to be guarded from contamination. It was only by threats of dire consequences that the Maharaja of Alwar, a particularly disagreeable representative of the princely order, was prevailed upon to shake hands with his monarch, George V. His Highness, who could produce an authenticated genealogy to establish that he was decended from the god Rama, had objected to shaking hands on the grounds that the touch of an infidel would damage the purity of his line.

Everything else was pretty much in conformity with this stance. Processions, military parades, durbars and other forms of pageantry left little time or money for hospitals or road-building, and the industrial age was kept at bay as something alien, an imposition from the West. The elephant-trainer and the court jester ranked higher than the state engineer, and the keeper of the harem was far more important than the prime minister. Education was like gunpowder, useful only in the right hands, dangerous in the wrong. A good citizen was one who accepted unquestioningly not only his master's absolute right to be his master, but also that divinity was involved somewhere. And towards that kind of citizen a ruler had obligations: he must not be allowed to die of starvation. But that was the beginning and the end of a ruler's responsibility as the protector of the poor. *Aham brahmasmi*, 'I am the Universe', a catch-phrase pulled out of context from the *Gita* sufficed to define their philosophy of life, and all else flowed from that concept. The only manner in which successive Maharajas of Jodhpur could think of providing a livelihood for their famine-stricken subjects was to set them building yet another palace, never a reservoir or a canal. To this day it is doubtful if the schools and hospitals in Jodhpur outnumber royal residences.

To be sure there were exceptions, but too few to make a difference to the general rule: only half a dozen princely states could be said to have been administered as well as the areas under British control. The point must be made that quality of administration, not education or economic progress, was the sole criterion of good government; it was the pride and boast of the British in India, the entire substance of the white man's burden. That states such as Travancore and Cochin had more literacy or that Mysore had carried electric power to its remotest villages did not necessarily mean that they had done more for their people.

It was not even as though this handful of good princes were regarded with any special favour by their masters. Indeed if – as the Nationalist leaders often alleged – the British did not actually encourage the princes to keep their subjects in a state of backwardness, they certainly gave that impression. As time passed, they became more and more indulgent towards the princes' misrule and enacted stringent laws to protect them from public criticism in their territories. Within the states themselves, of course, the mildest criticism resulted in an indefinite spell in jail, and public meetings, even those of a non-political nature, were almost unheard of.

If it was British policy to isolate the princes and thus discourage them from joining forces, it also prevented them from squabbling amongst themselves. Centuries-old feuds and jealousies manifested themselves in trivalities: the number of race-horses kept, or mistresses or Rolls Royces; in the lavishness of entertainment, in the measurements of tigers shot; in the acquisition of jewels, dancing girls, erotic paintings and above all, of the titles and honours accorded by the British.

There were as many as five hundred and sixty-two of these princely states scattered all over India, ruled by Rajas and Maharajas, a Nizam, a Maharana, a Jam, a Maharawal, a Mehtar, an Ackond, a Wali, and scores of Raos and Ranas. The Nizam, entitled to be addressed as 'His Exalted Highness', ruled over a territory larger than England and Scotland put together (82,000 square miles) and a population three times as great as that of Portugal, (17,000,000). He also enjoyed the right to mint his own coins and to print his own currency. At the other end of the scale were the petty 'chiefs' whose domains did not cover quite one square mile but who, nevertheless, possessed ruling powers.

Broadly speaking, the rating of prince was governed by his 'gun salute' or the number of guns fired in his honour on the occasions of his official visits to British India – private visits did not rate gun salutes. The highest ranking prince was entitled to twenty-one guns, and there were five of these. After that the salutes diminished in twos, to nineteen, seventeen and so on, down to nine. At that only about half the total number of rulers, perhaps three hundred in all, were entitled to gun salutes; the others had none. Most of these three hundred were also entitled to be addressed as 'His Highness', the exception being the aforesaid Nizam who was 'His Exalted Highness'. In addition, many of the princes possessed other titles and honours, Sanskrit, Persian and British, given to them or their forefathers by their Hindu, Mughal or British sovereigns. The British periodically bestowed new titles as well as honorary military ranks as marks of special appro-bation, and the competition for such recognition was always fierce. As a class too, the princes were extremely sensitive to any neglect on the part of others in using their full complement of titles, ranks and honours, and from one's own subjects such a lapse would have been regarded as a deliberate insult. At the same time, to remember all the titles of an important ruler was not easy and mistakes were inevitable. For instance, the correct manner of addressing the last ruler of Baroda was: Major-General His Highness, Farzand-i-Khas, Daulat-i-Inglisia, Sir Pratap Singh Gaekwar, Sena Khas Khel, Sham-sher Bahadur, Maharaja of Baroda, G.C.S.I., G.C.I.E.

The three hundred 'salute princes' among them ruled a third of the entire subcontinent and a hundred million people. The only thing that Their Highnesses had to guard against was the displeasure of their political superiors who, apart from being in a position to sit in judgement over them, were also respon-sible for bestowing titles and honours. And while a few of the officers of the Indian Political Service, (as a few of the officers of any service in any country) were known to accept or even demand bribes, the large majority shied away from such blatant attempts to win goodwill. But it was not long before the Princes discovered a gaping hole in the armour-plating: the British passion for blood sports. A Resident or a PA (Political Agent) who might send back with a stern little note of reproof a gold cigarette case slipped into his briefcase, would never dream of declining an invitation to shoot tiger or bison or to chase jackals on horseback or to spear wild boars. So the princely order set itself up as the provider of blood

94

sports to the representatives of what they carefully referred to as 'The Paramount Power'. They worked hard at it and spent freely. In the states that published annual budgets (many did not) the amount set aside for *shikar* far exceeded that ear-marked for public works or education. After years of trial and error, they specialised in their local species of wild life, and by the turn of the century had acquired an official stamp of approval by virtue of the fact that the Viceroy, or the Prince of Wales, or some other British dignitary had participated in their hunts or shoots. The ruler of Bharatpur, a minor state, had the quite unique privilege of hosting the Viceroy's annual duck shoot. Indore, Jaipur, Gwalior and a dozen others could be relied upon for a tiger at short notice, Mysore and Travan-core for elephant or bison; Jodhpur and Dewas ran pig-sticking meets every weekend and Kolhapur a year-round cheetah hunt. Sandgrouse (appropriately, of the variety known as 'Imperial'), snipe, partridge, quail, woodcock, pheasant (crested), all acquired vigorous partisans and the competition to draw off the Viceroy and the provincial gover-nors from their accustomed haunts became increasingly bitter. And yet no official fry was too small for the Princely nets; the youngest subaltern or Assistant Collector had merely to hint that he would be passing through or was thinking of taking his annual 'casual' leave for which he had no definite plans and the crested letter of invitation would arrive...for he might be a future political agent.

Over the years, the species developed its own distinguishing characteristics which might be set down as follows:

The Maharaja: A large brown raptor found in most parts of India; prefers to live in isolation; distinguishable from all similar birds by his brilliant plumage which he does not shed at any time of the year; collects bright stones; is polygamous; even though unfriendly to his own kind, utterly tame when approached by a white man. Unfortunately, the female of the species has never been seen, for the male hides all his wives behind thick curtains which are known as purdahs...

They certainly spared neither effort nor money, but the effect was not always what they had aimed at. Some visitors were impressed, some repelled, by the extravagance and

95

misdirected energy. Who but a Maharaja would think of escorting a royal train by relays of horsemen galloping along the track, or concoct so impractical a means of locomotion as the elephant bus shown to Lady Dufferin in Alwar: 'A gondola-like carriage, two stories high, drawn by four elephants, all covered with gorgeous gold-embroidered cloths.'

By the end of the first world war, the Princes had familiarised themselves with the technique; no hotelier could have conjured up better European food – at least not in India – no professional white hunter would lead a safari with so sure a touch. While ostentation remained, much of the crudeness had vanished, and if elephant carriages were still around, they were not paraded. Even so pampered a visitor as the Duke of Windsor who, as the Prince of Wales, toured India in 1921, was impressed. 'Among these Indian autocrats,' he wrote, 'I found a way of life, almost feudal and sometimes barbaric, that had persisted for centuries...and there I enjoyed Oriental hospitality and sport such as I imagined existed only in books.'

The Prince complained to his father, King George V, that the very tight security precautions did not allow him to see things for himself. All the same he observed, not without approval, how the princely states were 'impervious to the growing uproar in British India', and it is indeed doubtful if His Royal Highness would have enjoyed himself quite so much had he not also discovered that here 'Mahatma Gandhi's menacing influence disappeared.'

Feudal, barbaric, impervious...how accurately do these adjectives highlight the lineaments of princely India! The sinks left to drain the corruption of India had also become repositories of India's traditions and culture, where mediaeval customs were preserved as if in brine; where turbans were the normal headgear and hookahs the normal smoke; where men rode after pig and drank lustily and women were no more than a muffled tinkle of bangles behind trellised windows.

In these islands of tradition, past and present met fleetingly and, in unguarded moments, touched hands. Here too, in a mood of self-searching, the twentieth century could look upon the nineteenth, as E.M. Forster did on a visit to Chhatarpur, and comment, not without a touch of regret, that there were: 'No industries. no railways...in every direction I see delicate ruined shrines and graceful trees.'

Today Chhatarpur is a tourist's Mecca, for that is where the fabled Khajuraho temples are, a sort of *Kamasutra* and *Anangaranga* and other Indian erotica transfixed in stone. But even when Mr Forster was there princely India was beginning to acquire a new identity. To be sure, it still remained the hunting ground of British officials, but it was now being spoken of as the 'real' India; more interesting, more exotic, more artistically stimulating, than British India.

On the map this other India was seen as bright yellow splashes on the Empire-pink background, but the two colours represented more than political divisions, they marked two worlds, separated in time much more than in space. One was Kipling's India, the other an Oriental Ruritania. In one, stiff-upper-lipped civil servants sat sweating under swaying *punkahs* while, in the coolness of Simla, their memsahibs played havoc with the careers of young subalterns on short leave; in the other, bejewelled Indians sported among nude women in marble pools or abandoned themselves to marathon religious orgies; one of Turtons and Burtons and Mrs Hauksbee, the other of cave paintings and dancing girls and court intrigues. One was 'The Raj', the other princely India; and like an unmarried couple that had gone on living together all their lives, they had become pathetically dependent on one another even though love and passion had long since seeped from their relationship.

As the Congress agitation against British rule gathered momentum, the British realised that the staunchest supporters of the Raj were the princes, if only because without it, there would have been no princes; the latter on their part made it abundantly clear that they wanted the Raj to remain; that they preferred British rule to any kind of self-rule for India that they could foresee – and thereby they drew upon themselves not only the displeasure of the Congress leaders, but of the mass of the Indian people. Throughout the nineteen-thirties, while the British themselves were making a half-hearted effort to devise a formula for a transfer of power to Indian hands, the princes directed their energies and resources to defeating these efforts. They kept harping on the sanctity

97

of their 'treaty rights' and demanded impossible guarantees. The coming of the second world war enabled the British to shelve these efforts for its duration, but by that time, the princes, by sheer intransigence, had managed to extract from them the promise that none of their territories would be transferred to the control of a native government without the consent of its ruler. A farcical situation developed. Any major prince, or any group of the less important princes, could effectively block independence merely by withholding their consent to be a part of that India.

But these rulers were not India; and to the British it had become clear that India did not want them. By the end of the war, they had made up their minds to quit. The new Labour government in Britain was not prepared, as its leader Mr Attlee put it, to let the Indian princes act as 'a positive veto on advance'. In other words, Britain was not prepared to go on ruling India merely in order to maintain the princes in their positions. The treaties, the British Government announced, stood abrogated, and that paramountcy, on which the entire structure of princely India had rested, had lapsed. As far as Britain was concerned, the princely domains were so many independent and sovereign countries, and it was up to them to work out their relationship with free India.

'We are being treated as a sort of no man's child,' moaned the Chancellor of the Chamber of Princes.

'Independent states! Britain is attempting to plant hundreds of Ulsters in our country,' pronounced the Congress leaders. And as though to forestall any further intransigence on the part of the Princes, Mr Nehru issued an unmistakable threat: any princely state that opted to stay out of the new India, he declared, would be regarded as a 'hostile state' and would be so treated.

At the same time, Mr Nehru and his colleagues in the government proceeded to reassure the Princes that what they were being called upon to relinquish was what they had never enjoyed: an independent status. The Congress, they told the Princes, had no intention of swallowing up their domains – at least not without their consent. All that Their Highnesses were now required to do was to sign a document known as 'The Instrument of Accession', conceding to Free India control only over those items which in any case their British masters had always reserved for themselves, namely, foreign affairs, defence, and communications.

The Princes got the message. They rushed to New Delhi and signed the 'Instrument'. Even at the time, many of them must have felt, as the Nawab of Bhopal expressed it, like the oysters invited to attend the party of the walrus and the carpenter, but any doubts they might have entertained were allayed by the statement of the States Minister, Sardar Vallabhbhai Patel, who told them: 'I invited my friends the rulers of States in a spirit of friendliness and co-operation in a joint endeavour. . . The Congress is no enemy of the Princely Order.'

The Raj died when its keepers pulled out on August 15, 1947; Princely India did not long survive its creators. The moment independence came, the Congress Government launched a systematic drive to 'integrate' the princely domains. Within two years, the last of them had been absorbed. Except in one or two cases where force was used, this was achieved by friendly persuasion reinforced, now and then, by some judicious sabre-rattling; and there can be no doubt that this was a remarkable achievement, 'one of the dominant phases of India's history', said Mr Nehru in a mood of self-congratulation.

And rightly so, for 'Integration' was achieved not only without bloodshed but, by and large, in an atmosphere of friendliness; as though the princes were trying to make amends for their initial obduracy and the Congress, on its part, had decided to ignore their lapses. In exchange for these domains and their assets the Government agreed to pay the princes and their heirs annual pensions which were to be called 'Privy Purses', and to let them retain their private properties and to go on using their honorary titles and other symbols of their order, such as the flying of personal flags on their residences and cars, which were so dear to their hearts. These arrangements were to be permanent and hereditary, and in particular it was laid down that the Privy Purses could not, at any time, 'be increased or decreased for any reason whatsoever.'

The Queen's proclamation and the Indian Constitution;

99

whatever either had spared to the princes was guaranteed to be eternal. But if, under the British Crown, eternity did not last quite ninety years, under the Indians themselves it showed signs of running out much sooner. Special privileges have always been a favourite target of political agitators, and in a country where the possession of a sub-standard motor car is not only a sign of wealth but something of a public affront, the pensions and privileges of the ex-princes were elevated into the most vital political issue facing the country. Parties which twenty years ago framed India's Constitution and under it guaranteed certain fundamental rights as being 'unalienable' and 'unalterable' committed themselves to wrecking that Constitution so that the princes were no longer be protected by it. The signs were clear: and in 1971 Their Highnesses were reduced to *'Shris'* and what could not 'be increased or reduced for any reason whatsoever' altogether abolished.

The Raj and its creations, the princely states, have both vanished, and today both appear equally quaint and ana-chronistic. Kipling in a sun helmet is no less a figure of fun than the Maharaja of Alwar in his elephant bus. And which kind served his subjects better – a Turton or Burton who took special pains to ensure that his decisions were 'minutely just; inflexibly upright', and prided himself on the fact that the moment he left office he became 'unapproachable' to India, or the prince who ran a polo mallet into a stack of files and divided them into 'yes' and 'no' decisions but lived among his subjects and spoke their language and, even in his palace, remained far more accessible than a sahib in his bungalow? And who served India better? The white man who erected military barracks inside Delhi's Red Fort or the brown one who squandered his money in building a replica of the Alhambra for his personal use?

Oddly enough, as far as the tourist is concerned, even in the new, emerging, independent India, the 'real' India is still what was, not so long ago, Princely India. Though he carries no crested invitation card his packaged tour guarantees a night's halt in a Maharaja's palace which has been converted into a hotel thus allowing him a taste of the grandeur that was theirs. He has heard of erotic carvings and cave paintings and he has been told that he must see some of the temples in the

100

south. He does not mind paying an extra dollar for a ride on a caparisoned elephant, and he hopes to be able to photograph a dancing girl, preferably uncaparisoned. He certainly does not want to be taken around a steel mill or a fertiliser plant. And on his way to Ellora or Ajanta, or Belur, Halebid, Khajuraho, Hampi, Chittore, Mandu, Amber, Bundi or Kanya-Kumari, his eye may be caught by the insolent whiteness of some marble palace that has been pressed into service as a storage shed for grain, or by a flashily turbaned camelrider silhouetted against a flamingo sky, or a householder quietly sharpening the family sword for its annual worship; and he may recognise these to be what they are, signals from a vanishing world. And before his luxury bus takes the next turning, he may look back and wonder if something of it should not have been saved and whether there is still not time to save it, before the last elephant is sold to a circus and the last polo field converted into a seed farm.

Reminiscences of Rural India

Kamala Markandaya

It is a little bit different now. The villagers are briskly organised, in particular those whose villages border a trunk
road or bus route where trader-type sophistication rides in
on the exhaust fumes. They will offer you coffee or coca-cola,
or oil for your car, setting up stalls at top speed to sell you
crude toys or handwoven shoddy, like the beady trash with
which chunks of Africa were bought, in exchange for extravagant sums in cash. Only in these transactions, of course,
the indigenous inhabitants have the best of the bargain.
Which, considering how long they have been out in the cold,
is perhaps as it should be.

How cold that cold could be I – and I suspect many more
people than would now care to stand up and admit it – did not
know until a car broke down on a cross-country journey,
miles from anywhere, carrying a party of townspeople including me, who needed shelter for the night and so stumbled
upon this small village in the Deccan.

It was a forgotten place. No one would have suspected its
existence, but for the wisps of smoke one of our party had
spotted. Some twenty-five huts huddled together in a rough
clearing, hemmed in by hilly ridges and those curious hum-
mocks one sees in the South, boulder piled upon boulder. A
meandering footpath led to neglected, overgrown fields. Other
than this there was little sign of cultivation. The volcanic rock
soil, black from trap and granite, might have been as it was in
the beginning.

Strangely, every door was closed: strange in that villagers
are open people, as hospitable as it is in their capacity to be,
who have kept their sense of humanity intact through centuries when a similar sense atrophied or was absent in their
governors. But – closed doors, and the sensation that begins

somewhere between the shoulder-blades that you are being watched greeted us. This went on and on, while we in the manner of townspeople reacted to the situation by shuffling our feet and avoiding one another's eyes and anxieties. Then, mercifully one of those blank wooden doors creaked open, and an old woman emerged followed by a dozen other villagers who had been crammed into the hut. Presently all of them were out, the whole village assembled in that neglected dustbowl, and the curious thing was the look they had, not quite fright, but certainly an alertness, as if they were used to taking to boltholes.

To start with, no one would say very much. When tea had been brought, and we were sitting in the hut with the hot tumblers in our hands, we learnt why. For the past ten years, since the outbreak of the war, the village had been haunted by dacoits. Lack of access roads, its geographical isolation, and a shortage of police, combined to make it a perfect hideaway.

'They come when the constables are after them,' said the old woman, thin hands stroking the furrows in her face which, perhaps, ate deeper at each visit. 'They take what they want. Our grain, our animals, our cooking pots. You see we have very little left. We grew good crops once. Now no one wants to till the fields, they would only come more often to devour the fruit of our labour.'

'How do you live?'

'We manage.'

Their living had shrunk to the frailest proportions, barely sustaining life and that was all. Later I saw something of the margin-paring that goes on in most rural areas, but nothing quite equalled this.

'We thought you were dacoits,' said a villager, 'otherwise,' he added, anxious to remove any thoughts of inhospitality, 'we would not have closed our doors on you.'

'Do we look like dacoits?' we asked.

'We did not stop to look.' They laughed heartily, with that superb capacity that human beings have of wringing a joke from the most crushing circumstances.

I never came across another village like this one, though I did hear there were others. Since then, of course, the country

103

has been opened up, roads are more numerous, and administrators more alive to their responsibilities.

Liking sequels, I tried to trace this village many years later; it must have been around 1956. It had completely vanished, its identity absorbed into a complex of cooperative landholdings and new, neat housing. The old woman had gone, but new people were about and smiled, and the new doorways stood open.

The other occasion involving closed doors that I experienced was during the drama engineered by Tha-tha. He was a peppery old man whom I subsequently came to know and like, though the beginning was unpromising. At this first meeting he was ensconced behind a bristling palisade of split and partially overlapping bamboos, so that in effect all we could see of him were slivers of bare brown body between the interstices, and a pair of black, absolutely furious, eyes.

It was disconcerting. Our little mission, charged with providing amenities for the villagers, had not expected barricades. Whereas the violent Tha-tha, it appeared, could think only in terms of battle, and had concluded extensive preparations to resist eviction and a levy on his goods for debts and taxes.

These preparations were meticulous and comprehensive, as we discovered when eventually the old man – looking rather sheepish now, the glitter gone from his eyes – took down several stout bamboo poles and invited us to enter. In one corner of the miserable mud hut was a water pot, of impressive proportions, to sustain the Tha-tha family during the siege. On a shelf above, a sack of rice, a bundle of kindling, presumably to cook it, a few chillies and limes, a box of matches, a tin can of kerosene and a hurricane lantern. Along the wall a spade, a plough, a hoe, a sickle, all carefully cleaned and the metal parts oiled, a locked tin trunk, a few blackened, earthenware pots, a few brass vessels, a pile of rolls of matting with fraying ends and two small grubby pillows.

The entire worldly wealth of Tha-tha and family, snatched from field and courtyard and gathered under a thatched roof that needed mending, the whole of which, at auction, if the auctioneer was good and the bidding brisk, would have fetched something like five pounds.

Details from the 'frescoes in stone' at Mahabalipuram to the south of Madras city. The elephants are part of the world's largest bas-relief – 80 feet in length and an average of 20 feet in height.

It was astonishing.

Yes, said Tha-tha complacently, it was not bad was it, considering how little he had started with; and he crossed his painfully thin arms across his chest in modest pride. We said our astonishment was over how little, not how much: whereupon Tha-tha took umbrage again. We should visit his neighbours, he said, many of whom had not put by a tenth as much as he. We should go and see for ourselves.

Truth has this unpleasant way of jabbing below the belt. It makes one slightly resentful, as if it is not quite fair play to be subjected to it at all. Sitting well fed and clothed in that hut, it felt wretched to be told that this abject array represented the wealth of a man's whole lifetime.

Tha-tha's eyes continued to flash. They were an old man's eyes: sunk deep into the bone, the corneas bearing the milk-blue rims of age, the socket surrounding as wrinkled as if purse-strings had been drawn; but for all that you could almost hear the blades being sharpened behind them. There were still sparks to be struck from this one, as well were discovering, though in our inexperience we did not realise how rare it was.

Just one thing, resumed Tha-tha. He could do so much, but he could not compel rain. Three dry years could finish him, as one dry year finished lesser men and their crops and their livestock. Then there was only one thing left to do: gather in everything of value, summon one's family, pen one's animals, and put up the barricades. This he had done.

He seemed to speak completely rationally. The gap appeared only after we had thought about it a little, showing where something seemed slightly out of true.

Surely, we said, putting a finger on the spot, if one earned nothing one owed nothing? Surely such things could not happen in this new humane independent era that had just begun? Tha-tha sneered. He snarled, literally, like a dog, and blew through his teeth. We city dwellers, he said, could tell him all the fancy tales we liked. He would tell us what went on. It was like this. Each village was assessed for so much revenue and through flood and drought that level had to be maintained. If it fell they came and seized everything one had and sold it. It had happened to him as a young man, and in his middle years before he had learnt to resist. He was not going

op: *The only remaining shore temple at Mahabalipuram is believed to have* 105
en built by the Pallava King Rajasimha in the 7th century. Bottom: *Part of*
e Meenakshi temple complex in Madurai (S. India), the mother city of Tamil
erature.

to let it happen again in his old age.

We went away to find out more and returned to reassure him. Or tried. He refused to believe us. Or perhaps he could not.

Impasse. One very often reached impasse in the villages, either coming up against a taboo, like the community that shared a grove with a tribe of monkeys, preferring to suffer their depredations than to risk the ill fortune which would follow their removal, to ingrown fears like Tha-tha's. But they were more easily resolved, perhaps because one dealt with them one at a time, without the paralysing awareness of the waiting queue. So, now, a collection was taken. The money was pressed into Tha-tha's palm – nervously, anticipating a proud refusal. It was instantly accepted – the old man nurtured no false inhibitions.

Whereupon both sides relaxed. A daughter-in-law was summoned and made tea. Children came and we were told their names. Tha-tha's hostility had died; he was, basically, a man of great gentleness, an old man whose frailty one forgot when faced with his spirit.

After some days, cautiously, Tha-tha showed us the rest of his possessions: a small plot of wet land, a few coconut palms in which he owned a part share, the milch cow and its calf – nothing but a pathetic skin, this, crudely stuffed, but still enough to flush milk into its mother's udders; and the bullocks that drew his plough. These beasts were well past their prime, but Tha-tha was nevertheless very attached to them, and they to him. He called them by name, patting their flanks in an abstract, affectionate way, like partners in a happy and long-established marriage. We often saw him rubbing them down, or hand-feeding them carrots or tufts of young green grass. Yet he was cruel too, the same hand would bring the stick down on their haunches with a horrifying crack. I have never, either then or since, been able to reconcile these two sides of Tha-tha. Years later, a campaign was begun to persuade cultivators to sell their draught animals once they were past their peak. I believe it met with a good deal of resistance, and I have no doubt it was because of the same kind of feeling that existed between Tha-tha and his bullocks.

Before we moved on there was a speech in the village centre

106

which started as a vote of thanks for the amenities provided and ended in things being said which are commonly said after revolutions. Someone spoke of government of the people by the people, and someone else spelt out why it could never be as callous and unaware as before. Tha-tha was in the crowd that listened, though one could see his attention was not exactly riveted. When we asked him if he was convinced he shook his grizzled head. 'The sirkar,' he said, making a wry mouth, 'who can tell which way the animal will jump?'

Well, I hope he found that the animal jumped the right way for him. Uncertainty as to the nature of the animal plagued both sides. Perhaps the government suffered more, through understanding less, particularly during the British era, than the rural communities they governed.

The idea of South Indian peasants revolting strikes a bizarre note. Not merely because they are a mild people as people go, but because the organisation to unite and rouse them had not come to full strength. Yet at local levels involving local interests there has always been a tightly-knit solidarity which appears only at times of crisis.

Such a crisis began in a village to which, in our student days, my friends and I were occasional visitors, and ran a knife-edge course between farce and disaster.

The village concerned is a pretty hamlet not far from Madras, but well away from trunk routes and accessible only by a rough track deeply rutted by cart wheels. Its pattern is a recurrent one: paddy fields, a cluster of thatched huts, a grove of palmyra trees, narrow footpaths leading to patches of cultivation of bananas and corn. A feature is the tank, a land basin that empties during the dry season but rapidly fills during the monsoon rains, which irrigates the main area of cultivation. In the centre of this tank is an island of Barrington trees, whose roots can withstand prolonged water-logging, and whose thickly interlaced boughs and branches provide ideal nesting sites for water-birds. Here, for as long as anyone can remember, birds have come in their hundreds: storks, herons, egrets, spoonbills, cormorants, ibis, teal, and a dozen other species. From August, when the breeding season begins, to April, when the birds depart, the island teems with activity. Even to casual visitors it is an absorbing spectacle.

Such tanks with their flocks of birds, though not widely distributed, occur also in other rural districts in South India. Wherever found they are by long-standing tradition treated as sanctuaries; local culture and humanity combining with self-interest to protect and preserve the bird colonies. This self-interest runs deep: it is concerned not with such superficial activity as the attraction of tourists, but with the fertilisation of soil upon which present and future generations depend for their livelihood. The fertilising agents are the birds themselves, whose droppings fall in the tank and imbue its waters with rich manurial properties. Such green fields are not to be seen, cultivators will tell you, nor such crops, as those nourished by these waters.

To this village, one August, came two British soldiers from a battalion (this happened during the war) stationed near Madras. Their arrival was unplanned: they had not known about the village. They were out on a day's duck shooting, in the course of which they had chanced upon it.

Their excitement mounting at the sight of the incoming flight of birds, they announced their intention of shooting and called for coolie-porters to handle the unexpected bag. They then took up positions on the *bund*, an embankment constructed of mud and clay to increase the catchment potential of the tank.

The horrified headman, hurriedly summoned, remonstrated with them in his limited English, pointing out that the place was a sanctuary. The result of this harangue, according to him, was that he was pushed aside by the two men, who began firing into the air in the trigger-happy manner of soldiers in wartime. Perhaps they also hoped to intimidate the villagers, who had by now converged on the trouble spot in force.

Country people however are not easily intimidated when their ancient traditions are being violated. Their fear was not that a few score birds might be shot, but that the entire colony might take fright and abandon the area, with grievous consequences for the land. They took action and began stoning the Englishmen.

At this juncture we, who being visitors were the last to learn of the affray, came on the scene. It was an ugly situation. Tempers were high, stones were flying. The two men, who to their credit even during the worst moments did not turn their guns on the people, were already bruised and battered, but showed no sign of yielding. One of them lost his balance and fell, dragging his companion down with him. The two slithered down the embankment and landed in the tank.

At its shallow edges the depth of water in the tank seldom exceeds four or five feet, so the men were in no danger of drowning. There were however the excited villagers, who ran onto the *bund* waving sticks and poles and shouting as only a volatile Indian crowd can shout.

But the poles were dipped into the thick scum carefully: the men clung to them and were dragged to the bank.

We drove them back to town. They had cleaned themselves up, but smelt terribly. It was a silent journey, except when one of them, the more ruminative, said, 'It doesn't do to underestimate peasants.'

No, indeed.

Since my first book, with its background of rural living, was published, I have often been asked to describe a typical peasant. Despite my limited experience, my considered reply is that a description of the typical human being will do. Your peasant is Everyman. Those differences which are said to mark him, are not readily visible to me, with one exception: the stupefying degree of endurance and resignation of which he is capable. The world of pleasure and plenty has every right to claim a part in preserving the conditions that have developed such capacities, and may well die regretting them: a process which has already started. The fundamental mistake is to think that a peasant thinks differently from you.

Consider their board. In a peasant's household it consists of boiled rice, *dhal* (lentils), a few country-grown vegetables, tamarind, salt, sugar, tea and a spoonful or two of milk. The gourmet's delights that he hankers for, and occasionally gets, centre on eggs, curd, milk in liberal quantity, and sweetmeats for the children. Were he to secure these, would he then be satisfied? Not at all. Take the headman, who is basically the same peasant. His diet sheet includes these luxuries, but now he aspires to fruit, spices, butter, and ice-cream for his children. Wanting that bit more, an amelioration or improve-

109

ment in living conditions for one's children if not for oneself, happens to be a common factor in humanity. It is a failure in realism, if not rabid egotism, to imagine that a ceiling on wanting operates in a rural society, when it is demonstrably absent everywhere else.

Equally unfounded is the notion that rural areas are oases of peace, fostered no doubt by a countryside largely free of the clamour, fumes and odours of modern urban living. One has only to stay in a village a few weeks, to see the myth part company from the reality. The fact is that mental stress and mental illness are as rife in these areas as anywhere else.

These ailing minds are afforded relief, and sometimes a cure, by men and women variously called *Swami*, *Guru*, *Mata*, *Thai*, or other respectful titles, who are usually found one to each grouping of villages. The frame in which they work is very simple: the small courtyard of an unpretentious temple, a wayside shrine, the shade of a banyan tree.

The location of such places is not always easy to discover, though well known to the villagers, who will come, sometimes from long distances, to seek help for their troubled and tottering minds. A woman who worked for us took me to one such shrine together with her daughter who, she said, suffered from fits. The mother was a cheerful, sturdy character. The daughter turned out to be a thin, melancholy woman, who flinched when spoken to, turning away her face one side of which twitched feebly but incessantly.

On reaching the shrine she sat by herself, on the periphery of the small crowd which had gathered before the *Swami,* eyes cast down and hands lying motionless in her lap. Occasionally she gabbled loudly, accusing herself of appalling sin and crime, and saying she was possessed by evil spirits which led her to do wrong. Other and earlier societies would certainly have found her a ripe case for burning at the stake.

In fact, she had suffered intolerably. Of six children she had borne, the first five, all girls, had died in infancy. The sixth child, a boy, had lived for two years before succumbing to cholera. At this point the husband, no doubt taking his cue from these self-accusations, piled the blame for all these disasters, as well as for a failure of his crops, on his wife and abandoned her. About this time, not surprisingly, her mind

110

according to her mother 'began to play tricks', and she became convinced not only that she had murdered her own children but other children as well.

'Now she is much improved,' said the mother. 'You do not believe, but then you did not see what she was like before.'

Instant improvement, however, I did see.

The transformation began as soon as her name was called. The nervous shambling woman, who had not looked up once, got to her feet, glanced round the assembly, walked confidently through its ranks and sat down cross-legged in front of the *Swami*, having first bent and touched his feet in reverence.

The proceedings were short and absolutely simple.

The *Swami* – a young man, but already possessing the calm of someone who has seen through a good many illusions – laid his hands on her in blessing. He then asked her to tell him all that troubled her, which she did. After this he explained, generally, that such powers as he had came from God, to whom he must turn for strength, and he invited the audience to join him in a hymn. 'Hymn' is possibly a misnomer: what followed was a rhythmic chanting, a steady *Ram-Ram-Sita-Ram* accompanied by a lone drum and bells which jingled in time to its beat. This went on for about fifteen minutes, in the course of which I noticed that the *Swami* (and indeed some in the crowd) had begun to sway a little. His eyes were glazed, and rigors shot through his body which looked unnaturally stiff, as if rods were slotted under the skin. His voice was also changed. Normally rather thin in the typical Indian way, it had grown strong and sonorous and could be heard with great clarity above the massed chanting.

At the end of this time the *Swami* seemed to return to himself. His whole body relaxed, and his eyes, though still vestigially dazed, began focusing again. He looked tired and drained, but if anything more peaceful than before. A few minutes went by. He raised a hand, and the chanting died. Then he placed both hands on the woman's brow and said, gently but with authority, 'The evil spirits in you I now take upon myself.' At the same time he made one or two stroking movements, as if massaging away a headache.

That was all.

A few moments later the daughter joined her mother,

111

greeting me as if she had never seen me before. The nervous tic had vanished. She spoke lucidly, describing in a matter-of-fact if graphic way how it had been before ('as if a tarantula was gripping my mind') and how free and in control she felt now.

I wondered how long it would last. Some months later I asked, and was told that the fits came on her only very infrequently. More time passed. When I next enquired, I heard that she had set up home with a widower with whom she was living happily. One day finally the mother, her face lit with smiles, announced the birth of a child. At that point two years had elapsed since she had last had recourse to the *Swami*.

Are all cases resolved equally satisfactorily? Many say that they are, or at least that there is considerable amelioration. My own observation of what one might call mental health clinics – though there is a case for considering them on a higher plane – is, unfortunately, too limited to enable me to judge. Those I have been to have diverged widely in their results, as in the methods used. In some the *Swami* dissociates completely, by which I mean that the whole session is conducted while he appears to be in a state of trance. In others it is the supplicant who enters this state. Still others are conducted quite simply as prayer meetings.

Certain features, however, appear to be common. Almost invariably, intending supplicants refrain from sexual intercourse the day before. I once asked why, but no reason was forthcoming; it is an unwritten rule, spread by word of mouth. Invariably the *Swami* fasts, prior to a healing or praying session. Villagers who have suffered prolonged hunger will sometimes describe the floating sensations experienced, or visions fleetingly glimpsed. It would not be surprising if the mystic used some such aid for splitting away from worldly reality. Finally, the *Swami's* time and efforts are never directly matched to one's wealth. From gratitude, or reverence, people will bring gifts of fruit, flowers, food, money and even jewellery: but nothing is demanded, and often what is accepted is returned in provisions for the old and sick. In the West many of those concerned with healing, the doctors and the psychiatrists and the analysts, demand a standard of affluence which is apparently necessary to the status of their calling, and which often sets them apart from most of their patients. In contrast, in rural India it would be the exception to find a *Guru* practising mental healing who does not move among the poor, a poor man himself.

Some time ago there was an article in an English magazine about a man with powers of precognition. The reporter asked if he ever used them to enrich himself, say by backing the winning horse prior to the race. The man replied if he did the powers which had come to him as a gift would cease to work, and in any case he would have scant respect for himself if he tried.

I remember an Indian journalist saying something of the kind to a *Swami* he had tracked down. Why, he asked him did he choose to live in such poverty, existing on roots and leaves as if he were destitute, despite the gifts that were showered upon him? The *Swami* replied tolerantly if obliquely, pausing in his consumption of some cactus fruit which the village children had gathered and brought him. 'This way,' he said, displaying the sticky half-eaten fruit with the forbidding spines in their crimson middles, 'I taste the sweetness. But if I chose some other, my tongue might feel only the prickles.'

Rural India has provided me with a fair stock of salutary memories. This one sticks.

Religions of India

C.C. Joseph

India is the birthplace of many religions and, through the ages, Indians have manifestly tended towards a predominantly religious outlook. The average Indian's life is rigidly regulated by the taboos of his religion. The main ones that have developed in India are Hinduism, Jainism, Buddhism and Sikhism. Among those that found a congenial home in the subcontinent, Christianity, Zoroastrianism and Islam deserve special mention.

Hinduism

Embracing around 440 million people, about 83.4 per cent of the total population, Hinduism, which has persistently shown an extraordinary capacity to synthesise competing faiths into its own philosophical system, shelters under its umbrella monists, monotheists, polytheists, pantheists, animists, totemists, agnostics and even atheists.

Hindu religious thought started with the *Rig, Yajur, Sama*, and *Atharva Vedas* (books of knowledge) of the Aryans, between 1500 and 900 BC. Other important books of the Hindu scriptures are *Brahamanas* and *Aranyakas* (both dating from 600 BC); *Upanishads* (400–200 BC); *Puranas* (AD 500–1500), and the two great epics, the *Ramayana* (350 BC – AD 250) and the *Mahabharata* (200 BC – AD 500).

According to the most abstract philosophy of *Vedanta* (some of the *Vedas*), God is a transcendant being, an all-pervading Universal Spirit, unborn, imperishable, inconceivable, omniscient, omnipotent and omnipresent, known as *Parabrahma*, the Ultimate Principle, beyond which nothing exists or has any reality. *Parabrahma* is not worshipped and the great sages only meditate upon it. But at lower levels of religious practice there are those who pay homage to lesser gods, ghosts, demons, ancestors, the sky, the sun, the moon, the planets, stars, mountains, rivers, the cow, the monkey, the serpent, the rat, the male and female generative organs, stones and trees. The· worship of the serpent and the phallus are of particular interest. Serpents, it is believed, are supernatural and possess great mystical powers. The earth is supposed to rest on the heads of the thousand-headed Adisesha. The worship of the *Lingam,* the phallic emblem of God Siva, is particularly common in South India, although the *Rig Veda* condemns it.

All Hindus believe in the doctrine of *karma* or predestination. The soul is eternal and every soul is thought to be involved in a cycle of births and deaths. A man's destiny in one life is determined on the basis of his *prarabdha karma* or deeds, good or bad, committed in the previous life. A virtuous man is reborn to enjoy the fruits of his virtues, a sinner is destined to suffer for his vices. For the shaping of this cause-and-effect chain man alone is responsible; and man is punished by his sins, not for them. Hindu philosophy asserts, however, that man is actually one with the Real, the Ultimate Principle and thus every human being is a part of the Universal Spirit, with all its absolute attributes latent in him.

Hindu mythology has a pantheon of 330 million *devas,* or lesser gods, presided over by the great trinity: Brahma, the God of Creation; Vishnu, the God of Preservation, and Siva the God of Destruction. Four-headed Brahma is married to his own daughter Saraswati, the Goddess of Learning and Wisdom, and ancient cosmology traces his birth to Brahmanda, the cosmic egg, born of the Supreme Absolute. One day of Brahma is equivalent to 8,640 million human years and at the end of a hundred of his years the entire creation, together with Brahma himself, vanishes. Vishnu is a spiritual Gòd, and from his navel sprouts the lotus flower on which Brahma is seated. His vehicle is Garuda, the eagle, and his wife Lakshmi is the Goddess of Wealth and Bliss. Mother Earth is also one of his wives and Kama, the Cupid of the Hindu pantheon, is his son. Siva is associated with the malignant and the mysterious in nature, his great cosmic dance

115

typifying the ordered movement of the universe. He is also God of Regeneration and Sexuality and his consort, Parvati, is hailed as the personification of primeval energy.

Hindu mythology is full of the adventures, the clashes, the romances, the follies and the foibles of these gods and goddesses and other supernatural beings, such as demons, ogres and spirits. One story tells how Siva once caught Brahma, the Creator, telling a lie and rashly cut off one of his five heads. This was a cardinal sin, and because of it Siva found Brahma's skull stuck to his palm and had to perform an age-long penance; to wander as a mendicant, holding the skull as a begging bowl until the sin was expiated and the ugly appendage fell off. Even so his tarnished image as a beggar led to the greatest catastrophe in Siva's life. His first wife, Sati, humiliated, burnt herself to death and Siva retired to Mount Kailasa to spend years in austere asceticism. Meanwhile, Sati, reborn as Parvati, the daughter of the Himalayas, determined to rejoin Siva, but he would not look at her. Seeing this, Kama shot his arrow which had the desired effect: Siva opened his eyes and became enamoured of Parvati. But, provoked by Kama's interference, Siva opened his third eye on his forehead which acted as a death-ray and reduced him to ashes. Siva and Parvati had two sons, the six-headed Kartikeya, God of War, and the elephant-headed Ganapati, the divine remover of obstacles. Besides these two odd beings, Siva is credited with another son, Hari-Hara (Vishnu-Siva) whom he begat in his brother god Vishnu, after making the latter assume female form. This came as a happy climax to the churning of the Celestial Ocean of Milk. The story goes back to the dawn of time, when the 330 million *devas* and Indra obtained the help of their traditional enemies, the *Asuras* (demons) in the mighty task of churning the primeval ocean to obtain the divine drink of deathlessness, *Amrita*. As the pot containing the *Amrita* appeared on the surface of the ocean, the demons disputed the right of the gods to have the first quaff. According to one version, they actually made off with the pot across the sky, resting it or spilling it on the way at four spots, Allahabad, Hardwar, Ujjain and Nasik. In any case, a conference was called and Lord Vishnu appeared on the scene in the guise of a ravishingly beautiful damsel, Mohini (Enchantress). She made both parties agree that she should serve *Amrita* to all of them impartially if they handed over the pot to her and sat blindfolded. Mohini served the nectar to the *devas* first but then ran off with it, thus cheating the demons. When the story reached Siva he asked Vishnu to put on the Mohini form again for him to see, and as he did so, Siva embraced Mohini sexually thus producing Hari-Hara.

Vishnu came to earth in nine incarnations and is due once more, as horse-headed Kalki. First he came as a fish; then as a tortoise; thirdly as a boar, and fourth as a half-man, half-lion being. His next incarnation was in the guise of a mendicant dwarf, and his sixth appearance on earth was as Parasu-Rama, Rama with the Axe. His next incarnation was as Rama, the son of King Dasaratha of Kosala, in North India. His purpose was to destroy the demon king Ravana of Lanka (Ceylon) who was tyrannising the fourteen worlds of the Hindu universe. The dramatic fulfilment of this mission, described in the *Ramayana*, one of the world's greatest epics, came about after Ravana abducted Rama's consort, Sita, and Rama with the monkey hordes under Hanuman of the southern kingdom of Kishkindha, crossed over to Lanka and burnt it to ashes. When Vishnu came the eighth time it was as Balarama, the fair-skinned elder brother and constant companion of the ninth incarnation, dark-complexioned Krishna. Perhaps the most colourful of all the Vishnu incarnations was the last, as Krishna, the eternal lover and favourite of the milkmaids of Mathura, just south of Delhi. After killing off the demon king Naraka, he released from the latter's harem 16,000 women whom he married *en masse*. Krishna came primarily to slay the tyrant, Kamsa of Mathura, his own uncle. But he is best remembered for the support he gave to the five Pandava princes, who had been cheated of their kingdom of Hastinapura (ancient Delhi) by their hundred cousins, the Kauravas. The great battle fought at Kurukshetra to the north of Delhi, ended in the utter defeat and extermination of the Kauravas. The discourse Krishna gave to Arjuna, the Pandava archer prince, while serving as his charioteer on the battlefield, is enshrined in the *Bhagavad Gita* (or *Gita* for short), the Divine Song of the Hindu scriptures, and Krishna's sermon enlightened Arjuna on the path of man's duty and

117

the regulation of his thoughts and actions. It is the cream of Hindu philosophy and is venerated as the spoken words of God.

The poem dwells on the benefits of self-mortification and aceticism and thus puts its stamp on the practice of *yoga*. The aim of *yoga* is to merge the human soul with the universal soul. The chief forms of *yoga* are: *Karma-yoga*, for salvation through deeds; *Bhakti-yoga*, through faith; *Jnana-yoga*, through knowledge; *Mantra-yoga*, through chants; *Laya-yoga*, through activation of the subtle centres of the body, and *Hatha-yoga*, the best-known form, with its bodily postures, control of breathing, continence and meditation, which helps towards the attainment of *Samadhi* or super-consciousness.

Another unique feature of Hindu life is the caste system, which has the sanction of the *Gita*. The caste system is based on a *Rig Veda* passage which enjoins the Brahmins, who were born from the mouth of Brahma, to be the priestly class; the Kshatriyas, born from his arms, to be the warrior class; the Vaisayas, born from his thighs, to be cultivators and merchants, and the Sudras, born from his feet, to be menials. This in effect led to rigid class distinctions based on birth. Post-independence reform has abolished caste inequalities legally, but the rigours of the system linger on in remote areas.

Marriage is a must for Hindus, particularly on religious grounds. A departed soul has to be nourished and supported on its journey by the post-funeral obsequies performed by a male descendant. It was common for a man to take a second wife in case the first one failed to deliver a male child. Though not as popular as polygamy, polyandry was customary among many castes, including the royal clan of Pandavas. Divorce was not common, but Hindu law-givers did permit it in extreme cases. Modern law permits divorce. The ancient law also insisted on child marriage so that no girl remained without a husband when she attained puberty, but these are now banned.

Women of the Vedic period seem to have had equal status and rights with their menfolk. But they had become anathema by the time of Manu, the great law-giver (between 100 and 300 AD). Until adolescence they were placed under the tutelage of the father, after marriage under that of the husband, and as widows under that of their sons. Courtship was unknown and all marriages were arranged by parents. This continued until the middle of this century. Today, with property rights and career opportunities made equal for men and women under the law, the position of women in Indian society has vastly improved. No longer cloistered within the walls of the family, girls go to schools and colleges and qualify themselves for independent careers. Women have served new India as executives, legislators, governors of states, ambassadors, high court judges, cabinet ministers and even as prime minister.

The Hindu calendar is punctuated with perhaps more annual festivals than any other. The greatest of them, *Divali*, the festival of lights, is celebrated all over the country on five days in October-November. Originally a fertility festival, it now means different things to different people. *Ganesh Chaturthi*, in August-September, celebrates the birthday of the pot-bellied, elephant-faced Ganapati, the God of Luck. *Navaratri*, in October, re-enacts the life story of Rama and Sita and the destruction of Ravana the monster. *Holi*, a boisterous spring occasion, commemorates the frolics of Krishna and there is dancing in the streets and coloured water and powder are thrown over passers-by. *Janmashtami*, in August-September, marks the birth of Krishna, to which many miracles have been ascribed. *Siva-Ratri*, in January-February, is sacred to Siva and is observed with fasting and vigil on the preceding night, followed by festivities on the actual day. And *Rama Navani*, in March-April, celebrates the birth of Rama.

There are also the regional festivals, such as the *Kumbh Mela* (pot fête) which commemorates the spilling of the divine nectar, in Allahabad at the confluence of the Ganges and the Jumna; in Hardwar on the Ganges, in Ujjain on the Sipra, and in Nasik on the Godavari. In these places a *Kumbh Mela* is held every three years but the biggest *mela* is in Allahabad every twelve years. *Narali Purnima* or Coconut Day is celebrated in Bombay, when Varuna, Lord of the Oceans, is worshipped with chants and coconuts thrown into the sea; *Pongal* in Tamil Nadu, when· the cattle are garlanded and venerated, and *Onam* in Kerala.

Jainism

Jainism, the first reformist offshoot of Hinduism – which
figured as an antithesis to its emphasis on a Creator God –
assumed prominence under Vardhamana Mahavira, who
lived in Magadha in north-eastern India between 599 and 527
BC. It is claimed that it existed even earlier as the religion of
the *Mirgranthas* (those who are free from bonds).

Mahavira was an older contemporary of the Buddha. He
renounced the world at the age of thirty and is believed to have
received 'enlightenment' while meditating under an asoka
tree. Mahavira found no need for a Creator or First Cause.
The universe has existed through all eternity and will continue
to exist. Matter is eternal. The infinite changes in the world
are due to forces inherent in nature and not to any divine
interference. Mahavira preached the complete subjugation of
the senses to the spirit, regarded as the supreme knowledge.
According to Jain belief, the universe runs in immense alter-
nating cycles of advancement (*utsarpini*) and retrogression
(*avasarpini*), each of incalculable duration. One ascending and
one descending cycle make one age; and the process repeats
endlessly. Each cycle has twenty-four cardinal saints, known
as *Tirthankaras*, or patriarchs of the highest order; its twelve
universal emperors; and its sixty-three great men. Mahavira
was the last of the *Tirthankaras* to come and go in the current
cycle of retrogression, which is to continue for the next 40,000
years or so.

To the Jains every object, even metals and stones, have
souls, plants and trees having at least two souls each. Jains
are permitted to eat things having not more than two souls;
hence they may drink water and milk, also eat fruit, nuts and
vegetables. To eat animals, which have three souls, would
violate the cardinal law of *ahimsa* or non-injury, and is forbidden.

Like the Hindus, the Jains are also obsessed with the doctrine of rebirth; and every Jain is expected to work out his own
release from the cycle of births and deaths through good
intentions, right knowledge and correct conduct. The best
way to do this is to become an ascetic. Jain monks have their
hairs plucked out one by one; they wear a one-piece loin

120

The ornate capitol of the highly decorated stone column in the Hall of Privat
Audience at Fatehpur Sikri. This column served as a throne for the Empero
Akbar, who sat atop it when receiving ambassadors or nobles.

cloth, carry an alms bowl, a staff, a broom (to sweep the ground lest they tread upon and kill insects), and cover their mouths with a muslin mask to prevent insects entering by chance and perishing. The laymen also keep certain vows, refraining from falsehoods, stealth and sensual pleasures. Jainism in its purest form did not accept the caste system but the impact of the social environment brought class distinctions into its ranks in a vague form.

During the first century AD the Jains split into two sects: the white-clad *Svetambaras*, and the sky-clad (nude) *Digambaras*, who held women as the greatest curse of humanity.

There are over 2 million Jains in India today who are, on the whole, successful traders and bankers. The community runs retreats for orphans and widows, rest houses for pilgrims, and hospitals for birds and animals. Jain temples, usually constructed on hill tops, are among the most ornate in the country.

Buddhism

Buddhism, the reformist movement which overtook Hinduism like a whirlwind under the patronage of Emperor Asoka in the third century BC, was founded by Gautama Buddha (563–483 BC), a Kshatriya prince of Mongolian stock from the Indo-Nepal region, and has over 3 million adherents in India today. His teachings dominated the country until Sankara's Hindu revivalist movement in the eighth century AD won it back.

Buddha, like Mahavira, was opposed to ceremonial worship, religious rituals, sacrifices, the caste system with its Brahmin domination and the authority of the *Vedas*. But he did not believe in the rigid ascetic way of life as preached by the Jains. The sad things of life he observed as a youth, such as old age, sickness and death, led Prince Siddartha, as he was known before he became the Buddha, to ponder; and the tranquility he saw on the face of an ascetic filled him with a new hope. On a determined quest for enlightenment, the prince renounced the world and left his home, his wife and his new-born child. He practised the severest forms of penance

op: *The life-sized stone elephant which, with a lion and a bull, mount guard* 121
er the most famous of seven rathas *at Mahabalipuram.* Bottom: *Kanchipuram known as the 'city of a thousand temples'. Containing 124 shrines, it is one of e seven holy Hindu cities.*

and asceticism, but gave them up as futile. Wandering from place to place he reached Bodhgaya where, under a banyan tree, he obtained the supreme knowledge he sought at the end of seven weeks of meditation. His first sermon, the *Dharma-chakra-pravartana*, or setting in motion the wheel of the law, was delivered in the Deer Park at Sarnath, near Benares.

Honoured as Buddha, the Enlightened One, he preached the doctrine that there is no soul and that the so-called soul is a physical and mental aggregate of form, or the physical body, feelings, idea or understanding, will and pure consciousness. He taught a law of causality, according to which ignorance generates desire, and this in turn creates the bondage of the senses, bringing pain. All *karma* or action, good or bad, would result in rebirth. To escape from the endless belt of rebirth, one has to avoid the two great extremes, namely, the attachment to passion and worldly pleasures on the one hand, and the practice of self-mortification on the other. The mean between the two extremes is called the *Madhyama Pratipada* or the middle path. The eight-fold middle path comprises right views, right aspirations, right speech, right conduct, right livelihood, right effort, right mindfulness and right meditation. By this path one could reach the ultimate state known as *Nirvana*, or extinction, which is the highest objective of Buddhism.

Christianity

India has a Christian Church older than any other in the world, except the church in ancient Palestine. This is the Syrian Church of Kerala, in South India, believed to have been founded in AD 52 by St Thomas, one of the Disciples. An early Syriac apocryphal work, *The Acts of Judas Thomas*, says that after the Ascension of Jesus, the Apostles gathered in Jerusalem and cast lots to decide where each one was to preach the Gospel. India fell to the lot of Thomas, and he came to North India. Modern historians are inclined to think that after a long journey he landed at Muziris (now Cranganoor) where he converted large numbers of Hindus in the region. He built seven churches and ordained priests to look after the new faithful. Legend says that Thomas then proceeded overland to the east coast where he continued his mission. He is believed to have been martyred on a hill near modern Madras and buried at Mylapore beach to the south of Madras. An underground chapel in the nave of the present Roman Catholic Cathedral of San Thome is stated to be the Apostle's grave.

Later, two waves of Christian immigrants from the Middle East, one in 345 AD and the other in about 800 AD settled in Kerala. It was their integration with the local Christians and the adoption of the Syrian liturgy by the local church that gave the name 'Syrian Christians' to the entire community. Today the ancient church, having assumed the name Syrian Orthodox Church, continues as an autonomous Church under its Indian prelate, the *Catholicos*, with his headquarters at Kottayam, but still acknowledges the spiritual supremacy of the Patriarch of Antioch.

Still later Catholic and Protestant missions, which came in the wake of the Portuguese, Dutch and English occupations, converted many more Indians to Christianity in various parts of the country. Christians form only about 2.5 per cent (13 million) of the total population. With little political pull or influence, the Christians of India have made their mark in the social, educational, medical and philanthropic fields, running an impressive number of colleges, schools, hospitals, dispensaries, leprosy centres, orphanages and institutions for the deaf, the blind and the handicapped.

Sikhism

The Sikh religion is a militant offshoot of Hinduism which came into being as late as the fifteenth century, in the Punjab, where Hindus and Muslims had come into regular contact and frequently clashed. Sikhism started as a pacifist movement by Guru (Preceptor) Nanak (1494–1538), who had many Muslim as well as Hindu disciples; and was nurtured on the same lines by his five successors. It turned into a militant brotherhood under the Sixth Preceptor, Guru Govind. And the tenth and last Guru, Gobind Singh, who had a great

123

fascination and even veneration for the sword, gave the community its present militaristic form. The Bible of the Sikhs is known as *Granth Saheb*. It is treated with veneration and placed on a special altar in Sikh temples.

Nanak was profoundly influenced by a great contemporary religious leader, Kabir, who, while accepting the basic tenets of Hinduism and Islam, condemned the idolatrous practices of the Hindus and the sophistication of the Muslim theologians. Kabir's verses have found a place in the Sikh scripture.

The *Granth Saheb* is opposed to the caste system, the dominance of the Brahmins, and many other Hindu ways of life. Consisting of 3,384 hymns, it is about thrice the size of the *Rig Veda*.

One great distinction of Sikhism from other Hindu-based religions is that it rejects the doctrine of non-violence. Nor are there scruples about killing animals for food. A Sikh is not necessarily born; he is rather initiated into the religion. The baptism is performed by stirring some sweet water in an iron bowl with a double-edged dagger. With this water the initiate is anointed. After this he adds the suffix *Singh* (lion) to his name. He also wears the five *kakka* (the five 'k's), namely: *kesa*, long hair; *kachcha*, short trousers; *kada*, iron bangles; *kanga*, comb, and *kirpan*, sword.

Found mainly in the Punjab, Sikhism embraces over 9 million people.

Islam

Islam was first brought to India by Arab sailors trading in Indian spices, soon after the religion had taken a firm root in Arabia. Its advent was peaceful and it took its place in coastal Kerala alongside Judaism and Christianity, which had already been accepted hospitably by the Hindu rulers of the region. The early Muslims of Kerala were so much influenced by the customs and practices of the local Hindus that many of them ignored the Shariat law of inheritance and adopted the matrilineal system of devolution of property from mother to daughter, customary among their Hindu Nair neighbours.

In North India, however, where it imposed its military might over a divided Hindu nation, its approach was different. The first Muslims came to the region as conquerors and iconoclasts. But once they had settled in the country and made it their home, they patronised the art and culture of the land. And in the centuries that followed, particularly during the Mughal days, the Muslims gave the people a new civilisation which has left its impact on all aspects of Indian life. There are about 85 million Muslims in India today.

Zoroastrianism

Another colourful community which has contributed much to the mainstream of Indian life are the Zoroastrians of Bombay, commonly known as Parsis. Followers of the great Master Zarathushtra, or Zoroaster, who was born in Media (modern Iran) in about 660 BC, they are descendants of the Persian refugees who in the seventh and eighth centuries AD fled to India and were given sanctuary. Their Bible is *Zend Avesta*, a collection of the sayings of the Zoroaster, which is said to contain some of the noblest philosophic and metaphysical principles and a highly elevating ethical code.

The Zoroastrians venerate the elements of nature. Although they do not like to be described as fire worshippers, they propitiate fire in their temples where they kindle sandalwood sticks and perform ceremonies. Their dead are taken to the top of the Tower of Silence and left there to be eaten by vultures, since they do not wish to defile the earth, water, air or fire with decaying flesh.

These diverse religions co-exist and with their characteristic features add colour and variety to the kaleidoscope of Indian culture. They have their own little jealousies and differences; but they have in common made the Indians friendly and tolerant and prone to accept 'the will of God' without grumbling. On the debit side, Hinduism and its off-shoots, which believe in the *karma* theory, have made their devotees complacent and fatalistic in their attitude to life and its material problems.

Astrology and Palmistry

C.C. Joseph

From the days of the *Vedas* (1500 to 900 B.C.) *jyotisha*, or astrology, has always been close to the heart of Indian life; and Hindus have always consulted a fortune-teller for advice on day-to-day affairs.

The first impulse of any man with a particular worry is to see his astrologer. Thus the disappointed job-hunter, the anxious student, the depressed bachelor, the desperate litigant, the incurable invalid, the childless couple, all go to him.

This apart, tradition makes it obligatory to seek the services of a seer in respect of the twenty-one ceremonies, starting from the pre-natal period of the child, which orthodox Hindus are expected to perform on auspicious dates and times fixed according to strict astrological considerations.

The *muhurtams* (auspicious times) for naming the child, for putting it on solid foods, for beginning its schooling, for the boy's first hair-cut, for his religious initiation, his marriage, its consummation and for the wife's home-going for her first confinement, have all to be decided in consultation with a diviner of the divine will.

It is perhaps in regard to marriage that astrology plays the most vital role. Despite the influence of Western culture, even educated boys and girls leave the initiative of choosing their life partners to their parents. And when the parents negotiate the marriage, the first step they take is to collect the *janma patrikas*, or horoscopes, of the boy and the girl and commission an astrologer to discover how far they match and complement each other.

Hindu astrology has developed a series of tests based on the time of birth of the boy and the girl to assess their affinity in matters such as health, physical fitness, sex, marital loyalty and temperament. The system also claims to detect whether their union would be fruitful in longevity, wealth and children. In cases where the boy's horoscope warns of the early death of the girl he marries, the prescription is to marry him with a girl whose husband is likely to die early, so that the two negatives make one positive and so ensure long life for both.

Hindu divorce is rare, in proportion to the high population, due, it is proudly claimed, to the guidance of the stars.

There is a blanket ban against starting anything auspicious at certain times. For example, the solar and lunar eclipse times, the new moon, as also the eighth day after the new moon, are regarded as unlucky for weddings, applying for a new job, beginning a new career, building a house, digging a well, sowing seeds and ploughing the first furrow in the field.

Then there are times prescribed to start treatment when one is seriously ill. On the other hand, there are also signs to indicate if a certain disease is incurable, and thereby the futility of attempting to cure it.

Each birth in the world takes place under one or other of the 27 asterisms mentioned in the mystic books. And on this basis each person has his own birth-tree, birth-animal and birth-bird besides his birth-star. One's tree may be banyan, bamboo, lemon or pine; the animal may be horse, elephant, goat, snake, dog, cat or rat; and the bird, crow, fowl, hawk or peacock or others in the list. One is warned not to destroy one's birth-tree, nor injure the birth-animal or birth-bird. Also it pleases the stars if one nurtures them carefully.

The illiterate villager looks up to the stars to tell him when his sowing would bring him good crops and asks the *jyotishi*, or astrologer, (or *panditji*, as he is addressed) when he might buy a pair of bullocks. However, he is in good company: the sophisticated city-dweller, the man of letters, the intellectual, the scientist, the politician and the film star all seek the *panditji's* advice. There is a difference only in degree. The Brahmins of South India, who rank among the most intelligent communities in the country, almost to a man tend to condition their lives to synchronise with the light shown by the stars. Architects and engineers also seek the astrologer's approval before starting the construction of bridges, dams and steel plants; scientists, when opening laboratories and technical institutes; surgeons when fixing the operation day in difficult cases; film producers to begin the shooting of a new picture; politicians when they submit their nomination papers for elections; and chief ministers before deciding on their cabinets' oath-taking.

There is a special reason why the Hindus are great believers in astrology. For them astrology is no fad or fancy; it is a matter of conscience and religion. According to the sages, nothing that happens in the universe is accidental. Everything has a cause and every cause produces its effect. You reap what you sow. Thus a man's present action contains the seed of his future, and similarly his present life is a verdict on his past deeds. The past, the present and the future are the three links of a chain which may be called the law of *karma;* and this chain runs through the unending cycle of births and deaths.

If we can tell the plant from the seed and the seed from the plant, then we should be able to read the future of a person from his present action or *karma;* as also to read about his past actions, which have made him what he is. This is the rationale of Indian astrology.

The Hindus believe in an inherent order of the universe. The universe or *Brahmanda* is the result of a gradual development or evolution of the creative power inherent in the Primordial Being, the *Parabrahma.* Every human being is a manifestation of the *Parabrahma,* intimately linked with its other manifestations, including the planets, acting and reacting to their movements.

The horoscope indicates the operative part of a man's *prarabdha karma,* or past deeds, which, in other words, is a blue-print of his destiny for this life. The indications in the horoscope express themselves in events in a certain sequence in accordance with the movements of the planets. A man's destiny and the movements of the planets during his birth are so perfectly dove-tailed that his life from that moment is guided by the influence of the planets as revealed in the horoscope. But it is emphasised the planets only indicate destiny, they do not not cause it.

However, it is not suggested that human life is absolutely and irrevocably subordinated to predestination. Man has a free will. The main purpose of astrology itself is said to be to discover the effects of one's past *karma* and to mitigate

128

the bad in them by appropriate *karma* or action in this life to one's advantage.

There is a perpetual tug-of-war between destiny and free will; and our present actions are the combined results of these two forces. The ordinary man's free will is not strong; and his actions in this life will correspond largely to the indications given in his horoscope. But a person of superior will force, or high spiritual attainments, can change his destiny, although even here the general pattern would tend to conform to the horoscope.

The twelve signs of the zodiac, the twenty-seven asterisms (clusters of stars) and the nine planets, which exclude the trans-Saturnine bodies but include the Moon's nodes, form the basis of the Indian predictive art, the moment of birth being the pivot on which it turns.

Each planet has a causative quality of its own. Thus the Sun stands for the subject's personality and father; the Moon for his mother; and Mars for his courage. Then the different signs of the zodiac play their own individual roles. The sign that is rising in the east at the time of a man's birth is his *lagna* or ascendant sign, or 'first house'. This house is predictive of the man's physique, prowess and fame. To cite a few more, the 'second house' gives indications regarding his wealth and manner of talk; the seventh his wife and married life; and the tenth his occupation. They also boost or, alternatively, afflict, according to circumstances, the planets placed in their jurisdiction. Thus, the 'eleventh house', which stands for gains, makes any planet located in it beneficial, while the twelfth, which indicates losses, renders those placed in it powerless.

Although all the planets impart their influences along, it is said that at any given time, one of them dominates principally over a man's destiny, another next to it, still another next to that and so on. The actions and reactions of the planets *per se* and the varied influences the signs they occupy exert on them lead to endless permutations and combinations of their individual potentialities – projected on to this time schedule.

The end results of the intricate calculations involved in horoscopic prediction often proves rewarding. The assassination of Mahatma Gandhi in January, 1948, had been pointedly hinted at by a Bangalore astrologer in April, 1947. A Delhi

129

pandit correctly foretold that Mr Liaqat Ali Khan of Pakistan would be 'shot dead'. About March, 1964, a *pandit* from Trivandrum (Kerala) ventured to assert that Mr Jawaharlal Nehru was bound to die before June that year and that his end would be sudden. Mr Nehru died in May. The same man predicted in mid-1966 that Mrs Indira Gandhi would not be in Indian politics beyond September that year. This last prediction, however, turned out to be an unqualified failure.

Another branch of astrology, known as mundane astrology, deals with the fortunes of countries, assigns ruling signs to the various countries of the world and forecasts their fortunes just as in the case of human beings. Many predictions of wars, famines, epidemics, earthquakes and other events have been made with remarkable precision. One might, however, add that mundane astrologer's forebodings of worldwide catastrophies during February 1962, when eight planets clustered for a while in one sign of the zodiac, flopped – to their dismay and humanity's relief.

Horoscopy usually starts with calculations made from the date and time of birth of persons actually born. But there is an interesting branch of astrology known variously as *Nadi, Bhrigu-Samhita* and *Satya-Samhita*, with vast libraries of horoscopes of persons written many years ago. Inscribed on palm leaves, these ancient manuscript volumes include the life stories of many people who were born only centuries later, and of many more yet to be born. They are believed to be the work of experts of a bygone age who marked all possible moments of human birth for centuries to come and wrote the horoscopes of the men and women who would be born at these moments. A well-known Indian writer has recorded that he was present when a *Bhrigu* astrologer read out from his collections a leaf which predicted a few days before the event, the death of Lal Bahadur Shastri, the then Indian prime minister, at Tashkent.

Another branch of fortune-telling, which is preferred for instant prophecies, is known as *Prasna Marga*, the 'question-method'. The *Prasna* practitioner sits ready with a picture of the ruling planetary position in his mind. The time of consultation, the first letter uttered and some other indication enable him to locate the client in his mental chart, and he

130

begins his calculations with the aid of a variety of small, uniformly-sized sea-shells called cowries. The questions asked of the *Prasna* practitioner generally concern the chances of getting a job, recovering a stolen article, success in a marriage negotiation, winning a law suit, and so on.

A somewhat refined version of *Prasna Marga* is *Doota Lakshana*, or messenger-reading, in which the messenger, or client, need not utter a word. This was once regularly practised by eminent practitioners of Indian medicine. The physician takes note of the hour, day, month and the lunar asterism and makes the astrological reckoning and fixes the messenger in his mental chart on the basis of the direction from which he came. He also observes the clothing, speech, caste, sex, mental state (excitement, fright, despair) and manner of arrival (running, walking, riding) of the messenger. For example, it would be inauspicious if during a particular asterism, the messenger were a Brahmin, a eunuch or a woman; or if he or she came at midday, midnight, running, or riding a buffalo; and if the physician were at that time asleep unclothed.

Significance is attached to every movement of the messenger. If he or she comes and stops in one place and then suddenly moves to another place in a particular direction, it indicates the person has come after visiting somebody else on the same mission. If he or she puts a finger in the mouth, nose or ear, the message contains bad news.

The foretelling practised by those physicians of Kerala who specialise in snake-bite cures is even more sophisticated. They mysteriously appear to know in advance when someone has been bitten by a snake and wait in anticipation for the case. Even before the patient or a messenger arrives, he can say which species of snake has been involved, how many wounds have been inflicted and in which part of the body, whether the bite took place inside the home or outside, and whether a cure is possible. In most cases he tells the patient where the snake could be found, but only with the strict injunction that it should not be killed in any event.

Almost as popular as horoscopy is fortune-telling by *Hastarekha Sastra*, or palmistry. Throughout India one comes across more palmists than astrologers.

131

The palmist shares the astrologer's belief in the doctrines of *Karma* and Destiny. But his faith in free will is better fortified than the astrologer's. For the palmist relies more on the indications of a man's right hand, which show what the man is actually going to experience as a result of his *karma* or free will in this life, than on those of the left hand, which depict his destiny from the previous birth.

In addition to the size, shape and skin-texture of the hand, the nature, length and directions of the lines, the prominence of the mounts, the setting of the thumb, the formation of the fingers and the shape and colour of the nails, Indian palmists take note of such non-linear indications as the conch, wheel, bow, arrow, fish, trident, lotus, bell, chariot, Sun, Moon, swastika, tortoise and bird, which they can detect on the palm.

It is claimed that the margin of error is far less in palmistry than in horoscopy, which could go off the tangent due to an error in the exact time of birth reported, and in which the relative strength of the planets has to be decided after intricate calculations, big enough for a computer to tackle. The palm is an open, fool-proof record, which leaves nothing in doubt.

Apart from the claims, however, neither astrology nor palmistry has proved more accurate than the other. Both have yielded astonishingly correct results, as well as miserable failures.

There is one form of palmistry which has a close affinity to astrology. This is 'thumb-reading', where the seer studies the lines of the thumb for the planetary positions and makes a chart of them. The thumb-reader claims all information about a man and his fortunes are stored in the thumb itself, so that there is no need to study the entire palm. The thumb is called the tiny replica of the total personality.

Thumb-reading, popular in South India, is based on the *Ravana Samhita*, named after the demon king of Lanka (Ceylon). The art is said to be particularly successful in detecting internal troubles and mental diseases. It has a curious method of assessing the status of people in terms of numbers. If a man's number is less than 5,000, he belongs to the lowest, struggling masses; if it is between 5,001 and 17,000, it denotes an average man of limited means and uneventful life; and if it is between 17,001 and 100,000, he belongs to a higher class with greater achievements. Astounding predictions through thumb-reading have been claimed.

Another set of experts, mostly in South India again, tell fortunes from *Samudrika Sastra*, or the science of reading the limbs of the body. This is a complete study of the different qualities of each part of the body from head to toe, including even the lines of the sole of the foot. Some of its practitioners take one look at a man's forehead, tell the asterism under which he was born and then read his fortune as if from a book.

The length of a person's shadow at a certain time of the day is used in reading his or her fortune by *Chhaya Sastries*, or shadow-readers, an art practised in Gujarat. It has its limitations, depending as it does on the availability of the sun and the shortness of the time during which the reading can be made.

Then there is the miracle man who helps you to trace stolen property or to locate a lost person or animal. He prepares a drop of ink by mixing soot in olive oil. Then he asks a child to gaze into the ink and describe what he sees. Children who have participated in the divination have stated how they could see the lost things.

But the most baffling of fortune-tellers are those that rely merely on their intuition and on no calculations of any sort. A blind mason-turned-fortune-teller, who claimed no knowledge of any system, was told two people had come to see him. After meditating for a few moments, he said one of them had come to consult him about a gold necklace his sister had lost. But it was his companion's brother who had stolen it. He said the thief had an accomplice. The two had earlier removed the offertory chest of a church. Then he predicted that the accomplice would run away and die in a distant place. The right hand of the thief, which had stolen God's property, would be cut off. And the necklace would be recovered.

Because there was no evidence, the police did not proceed with the case. But soon the alleged accomplice left home and later died far away. The thief one day went to the river to fish. He held dynamite in his hand, ignited the fuse but, being drunk, held it too long It exploded and the hand had to be amputated at the wrist. But the necklace was never traced.

Indian Sculpture and Painting

Mulk Raj Anand

In common with the artistic expression of all early civilisations, Indian art evolved to satisfy spiritual needs after material necessities, such as food and shelter, had been met. In the earliest known culture in India, the Proto Dravidian, in Mohenjodaro, Harappa, Rupar, Lothal, Maheshwar and Navdatoli there is evidence of this pattern of development. Following on basic social organisation, certain symbols, ideograms and images appeared – embossed seals with geometric shapes on clay pots, terracotta figures of a god similar to the future Siva, and bronze and stone idols. No unified system of belief is indicated – rather a kind of animism through which everything is ascribed to a spirit, a power. The emphatic female pudenda show an obsession with birth, fertility and procreation; the trident of the Siva figure indicates the awareness of power, the stance of the dancer the appreciation of rhythmic movement and the belly of the Harappa-type torso the love of volume and rotundity.

This primitivism characterises the terracotta remains of the next thousand years of Indian culture. Fertility images become expressions of the worship of the mother goddess, with big hips, protuberant breasts and pudenda even more emphatic than before. Female figures are more frequent than male and dryads, fauns, bird and tree spirits become elaborated into a pantheon of the forest life.

The Aryan conquerors, who probably destroyed the Indian valley civilisation as evidenced in Mohenjodaro-Harappa civilisation (c. 1500 BC), brought with them a culture which defied the elements and their manifestations. As they intermarried with the Dravidian population, which they enslaved, they absorbed their fertility images as well as tree spirits, snake gods, ghosts and hobgoblins. They decorated their fire altars with abstract geometric designs, but did not use clay, stone or paint.

In the period of synthesis between Aryan and Dravidian cultures, a single cosmic philosophy was created: once there was Brahma, the Supreme One, the Creator; his union with his consort created the manifold universe, but the ultimate objective is to return to the One. In practice, this view of the cosmos was expressed in prayer to, and contemplation of, certain symbols, such as the lotus, rising on its stem-axis above the waters of the universe, its petals open in all directions. The female pudenda of the surviving terracotta figures show that the female organs were worshipped as the source of life and the phallus, in the form of oblong stones, is exalted as the creator. Daubs of red paint on any wayside mound symbolised magical sacrifices to appease dead gods.

Some giant sculptures survive from pre-Christian eras. These seem to be funerary statues of kings, huge in volume; heavy, brooding and intense. The ghosts of kings may have been worshipped as in earlier Egyptian, Hittite, Assyrian, Babylonian and Achaemenid civilisations. Achaemenid influence in India survives in the tall Persepolitan pillars, topped by lions, which Asoka caused to be erected in various places after his conversion to Buddhism, carved with his edicts proclaiming peace, goodwill and righteousness. The buxom Yakshi of Didarganj has a shiny surface known as 'Mauryan' polish, which derived directly from the practice of burnishing surfaces common to the sculpture of the Achaemenid Empire. Between the 6th and the 3rd century BC this Empire, which had its capital at Persepolis in Persia, extended from The Nile to The Indus.

The rigid caste system imposed by the Aryans on the Dravidian population had lost none of its harshness in the merging of the two cultures. A series of revolts against this system culminated in Gautama's rejection of Vedic Hinduism and his sermon at Sarnath on attaining Buddhahood (Enlightenment) attracted a great following among the oppressed. But Buddha forbade images of himself, believing that art forms promoted physical desire, which causes attachment to life on earth, thus distracting from the striving towards *Nirvana* (release from the cycle of birth and rebirth).

After his death, however, the old need for concrete imagery reasserted itself and stories of his birth and various incarnations, from animal to human and Bodhisattva (potential Buddha), were transcribed in stone near the burial mounds (*stupas*) in which his relics lay.

During the 2nd century BC onwards, an expressive art of relief sculpture developed around the *stupas;* in Barhut and Mathura in North India, Sanchi in Central India and Amravati and Nagarjunakonda in the Deccan. This was a narrative art, which assimilated the vitality of the early terracottas, the preference for curves and volume and a primitivist depiction of dramatic scenes. These demonstrated the unity of all life; plant, animal and human; and the rectangular panels of reliefs, as in Barhut, or medallions, as in Amravati, show superb skill at constructing a parallel world of creative imagination in which legendary human beings live and move in intricate compositions.

At the same time as the gateways and outer walls of the Buddhist grave mounds were being decorated, a series of cave temples were being hewn from the rock in an arc-shaped area extending from the western ghat mountains into the Deccan. Speculation has it that these holy places served as resting places for traders (Roman coins have been found on the sites) as well as shrines for pilgrims. The walls of these temples, and their relief sculptures, were painted, and once again the stories surrounding Buddha supplied the main themes. The extraordinary skill with which the large-scale frescoes with crowded scenes were executed, has produced some of the finest art anywhere in the world as witness the temples in Ajanta and Bagh. The technique of painting perfected here later spread to Sittanvassal and Badami in South India, to Sigiriya in Ceylon, to Tsparang in western Tibet and Tun Huang in China.

In the north, after Alexander the Great's infiltration to the Indus, Buddhist sculpture flourished in Gandhara, first under the Greek rulers and subsequently under conquerors from Kushan in western China. The hybrid art which developed in this period resulted in Buddha images modelled on the Greek god Apollo. It is probable that this fashion spread to Mathura under the Kushans, for the emperor Kaniska, who ruled a vast empire, including Gandhara, from three capitals – Kapisa in

136

The dance of the devadasis, *Bharata Natyam, is prevalent mainly in south-ea India. The performer here is Indrani, author of the relevant chapter in this book*

INDIAN SCULPTURE AND PAINTING

Afghanistan, Peshawar on the north-western frontier of India, and Mathura on the River Jumna – adopted Buddhism. The Indian craftsmen of Mathura certainly improved upon the Gandhara Buddhas; for in the Jumna and the Ganges river basins portraiture was inspired by mental images rather than physical reality, the ideal image of the Buddha being defined in the holy text as 'round, round, seven times round'. The Yogi figure, seated in contemplation, progressed from the Katra Buddha in Mathura, with its sharply angular arms and legs, to the benevolent standing Buddha of Mathura, (5th century AD), the bronze Buddha of Sultanganj (6th century AD), and, most graceful of all, the Sarnath Buddha (5th century AD). This progression in Buddhist art influenced Hindu sculpture during the uneasy coexistence of the two main faiths in the classical renaissance of the Guptas from the third to the fifth centuries AD.

One of the greatest works of world art, both in size and quality, which can be seen at Mahabalipuram, near modern Madras, is of the 6th century AD. Commissioned by King Mahendravarman to inspire his armies, it depicts the episode from the Hindu epic *Bhagavad Gita* about the five Pandava princes and their hundred wicked Kaurava cousins (see Religions of India p. 117). Carved as a giant relief into a wayside rock it contains human and animal figures in dynamic movement and has the coherence of the crowd scenes in the Ajanta paintings executed four hundred years earlier. Mahendravarman's son, Mammala, a devout Hindu, extended the tradition of the Mahabalipuram rock to the shore temples in Mahabalipuram, in which the human figures, both female and male, are conceived in cylindrical forms through which the master craftsmen achieved a new fluidity of movement. These large-scale carvings continued to be executed until the end of the 6th century, culminating in the relief cut into the rock in the Durga cave in Mahabalipuram, depicting Vishnu asleep.

The conflict between the two faiths seems to have intensified the desire of the priests of both to seek artistic patronage. The Hindus appear to have been more successful, and from the end of the 6th century we get carvings of the gods from the Hindu pantheon which excel previous achievement in virtuosity. The pink sandstone Vishnu in Mathura (6th century

137

op: The eastern gateway to the Great Stupa at Sanchi, built by the Emperor soka. Bottom: An 11th-12th century figure of Buddha in Tejpala Temple, Dilwara, in Rajasthan.

AD) was foreshadowed by the standing sandstone Buddha, also in Mathura, which was carved at the end of the 5th century AD. The recumbent Vishnu in Deogarh, Central India, (6th Century) is reminiscent of the sleeping Vishnu in the Durga cave, but is an improvement on the latter in its harmonious handling of volume. The *Nar Narayana* relief in Deogarh (6th century) is lively enough, but static in comparison with the *Marriage of Siva and Parvati* panel in Elephanta (6th century AD) and the Siva reliefs in the Kailasa temple at Ellora (7th century). The Mahabalipuram rock is more like a rehearsal when faced with the accomplishment of later monumental reliefs. The greatest achievement of the early mediaeval period of India is the temple at Ellora, hewn on three sides from solid rock.

During the 7th and 8th centuries, the Arabs invaded Sind and Gujarat and the wall-painting tradition waned. But even with the Huns' incursions of the late 6th century, in the north, and the Arabs in the west, temple-building continued in the east, especially in Orissa, and the principles of architecture, sculpture and iconography were formulated in the *Silpashastra* – handbooks which conserved the traditions of the creative arts.

In the great sun temple of Konarak, the Black Pagoda, in south Orissa, a structure surpassed only by the great Buddhist temple at Borobudur in Central Java and the Angkor Wat complex in Cambodia, there are colossal free-standing sculptures of musicians, dancers, elephants and horses, as well as reliefs illustrating battle scenes and the drama of the union between male and female. The imaginative construction of the whole is beautifully realised by the integration of the sculpture and the architecture – the movement of the figures giving energy to the static rocks which surround them. In Khajuraho, this principle was carried further. Here architecture is sculpture and sculpture is architecture. The integral unity of the design is achieved by grouping the small conical *sikharas* into the texture of the big *sikhara*. The vivification of the whole is achieved by the use of the parable of creation through love scenes of the most naked but tender character. The rhythmic intertwining of male and female forms celebrates sexual union made into the actual experience of loving as a sacred rite. Nowhere in the world has such grace been infused into the union of male and female as in some of the panels of the Kandariya Mahadev temple. The carving achieves a quality of finish beside which all precedents seem to be mere shadows.

The itinerant craftsmen of northern India were doubtless involved in the building of the Siva temples in Rajasthan and the style of Khajuraho is obvious in much of the work in the Neelkanth Mahadeva temple in the forests near Alwar. A similar phase of constructivism flourished under the patronage of the Jain monarchs of Gujarat between the 11th and 13th centuries. On Mount Abu, they built a temple of white marble, transporting the stone from the plains below. It was carved using the technique of the ivory carvers, achieving fine details on the pillars and pediments which support the ceilings. The work here is unique, because the decorations are not chipped, but scraped, with incredible mastery of detail.

The grandiosity of the temple sculptures was balanced by several masterpieces created by the *cire perdu* (lost wax) bronze technique, examples of which are the 6th century standing Sultanganj Buddha (now in the Birmingham Museum in the United Kingdom) and the Brahma from Mirpur Khas in Sind in the Karachi Museum. This method of casting bronze was quite widespread in Kashmir, in the Pala empire in eastern India and in Chamba in the Punjab Himalayas. In south India, this tradition was particularly encouraged under the Chola dynasty between the 10th and 13th centuries. Some of the best examples of this technique are of Siva as Nataraj, Lord of the Dance. The iconography of Siva as dancer embodies the Hindu concept of rhythm – he is shown stamping the dwarf Evil under his right foot, with left leg upraised, arms poised in various gestures, inscrutable face surrounded by the symbolic circle of fire of the universe – transfixed in motion. Rhythm itself, as expressed in the controlled but violent movement of the dance, radiates from the best of these bronzes. Some of these are now in European museums, but some of the finest are in the Madras Museum. There are other important bronzes, modelled by craftsmen of genius, and the female Parvatis, the Somaskanda compositions and the portraits of the saints should be mentioned in this context. The *cire perdu*

139

technique is still practised in south India, though a decline of faith has led to the diminution of creative fervour since the end of the mediaeval period at the end of the 17th century.

The itinerant court of the Muslim Sultans in the 12th to the 14th centuries AD showed hardly any interest in the arts – apart from the building of mosques and mausoleums, though naturally enough there was a fusion during this period between native Indian art styles and imported Persian ones. The *Nimat Nama*, a cookery book produced under the Khilji monarch Sultan Nasir-ud-Din illustrates this synthesis. The rounded lines and bright colours of the figures in a Jain holy book, the *Kalpasutra*, are combined with the intricate foliage and rectangular frames of the Persian decorative style.

While this fusion of styles was taking place, the native tradition of the Jain holy books flourished along with the folk-inspired court art of those Hindu princes who were still independent, as in Mewar and some other parts of Rajasthan.

Although the impact of Persian art began to be felt in the 14th century Sultanate courts, the real challenge came in the 16th century when Humayun, the son of Babur, (the first Mughal emperor of India) returned from exile at the Shiraz court of Shah Tehmasp of Persia. He brought with him two Persian painters, Mir Syed Ali and Khwaja Abu Sammad, and commissioned them to give lessons to his thirteen-year-old son, Prince Akbar. The latter inherited his father's throne at the age of fourteen and spent the first few years of his reign in building an empire from his modest inheritance. During these busy years he had little time to spare for the creative arts, though he patronised craftsmen of all kinds and kept in touch with his Persian teachers, encouraging them to continue with a project, outlined by his father, of illustrating the *Dastan-e-Amir Hamza*, (the legendary tales of Islam) on large pieces of cloth. This innovation of painting wall hangings of cloth may have been inspired by Akbar's admiration for the murals in Rajasthan and elsewhere. In pursuit of his policy of achieving in his court a fusion of Muslim and Hindu elements, he attracted to his *atelier* native craftsmen like Basawan, Dashwant, Kesu and others and apprenticed them to the two Persian masters. In Basawan's *Tutinama* pictures, now in the Metropolitan Museum in New York, the fish-eyed, lionchested men and exhuberant females, animals and plants of the native tradition are constrained within the Persian fantasy framework. In the later *Hamza Nama*, the Persian figurative style is mixed with the density of traditional Hindu colours, and in one picture Rajput women are introduced quite arbitrarily into the background of a fabulous *Hamza* theme. Akbar took his painters with him on his expeditions to various parts of the country, and on his hunting trips. This resulted in depictions of the action they witnessed and was the beginning of the Akbaride style, with its emphasis on the exaltation of heroic action. He visited his *atelier* weekly, inspecting the work in progress, rewarding excellence and encouraging talent, much to the disapproval of the orthodox Muslim priests. He commissioned his artists to illustrate the Persian translation of the Hindu epic *Ramayana*, the *Razam Nama*. The Hindu artists soon outnumbered the Muslims in the workshop and, as Abdul Fazal, the Emperor's biographer, remarked, 'their pictures surpass our conception of things, few indeed in the whole world are found to equal them'. A visit to Cambay on the Gujarat coast in 1575 gave Akbar his first acquaintance with European painting and his artists became interested in the three-dimensional perspectives of western art. The introduction of western-style figures became fashionable, and when the Jesuit mission came to Fatehpur Sikri, five years later, they found that Akbar already had portrayals of Jesus and Mary decorating his dining room. The Jesuits presented him with a copy of Plantin's Bible engraved by Flemish artists, and Akbar set his artists to copy them. Thus the perspective of background landscapes which characterised European painting became a factor in Akbaride art; drapery was modelled; clouds, trees, houses, became more realistic, less stylised; and human beings became individualised when compared with the stereotypes of earlier Indian art.

Akbar's son, Jehangir, succeeded his father in 1605 AD. He had a genuine appreciation of painting, and could recognise the individual work of each painter in his *atelier*. In addition to the artists he inherited, he recruited Farukh Beg Kalmak from Uzbekistan, and Aga Raza and his son Abu Hassan from Herat; and Goverdhan, Bishandas, Manohar, Mansur and Muhammed Nadir flourished under his patronage. Jehangir, who had inherited a ready-made Empire, was by temperament a hedonist. He fell in love with the wife of the Governor of Bengal and wooed her until she married him. He initiated the custom of spending summers in Kashmir and became interested in landscape gardening, as witness the Shalimar and Nishat Bag gardens, near Srinagar. His vanity was evidenced by the numerous portraits he had painted of himself, in various moods and situations. His fascination with nature resulted in Mansur's studies of birds and animals, of which the turkey-cock, zebra and chameleon are justly famous for their accuracy and charm. Catholic in taste and interest, he wrote his memoirs and increased his collection of European paintings. He ordered his artists to introduce the western halo into their portrayals of himself and the divines. The three-dimensional perspective became more common and the European fashion of concentrating on one character in a work, subordinating other elements, was emphasised. One of the finest drawings of this period was of the death of the nobleman, Inayat Khan.

Jehangir's son, Shah Jehan, came to the throne in 1628 and remained there for the next thirty years. He had been a connoisseur of painting in his youth, but his travels both with his father and grandfather had inspired him with a special affection for architecture. His favourite materials were white marble and red sandstone, and he was responsible for both the Red Fort in Delhi – the Shahjehanabad – and most famous of all, the Taj Mahal, a monument to the memory of his wife. Shah Jehan indulged in his taste for the finest architecture. Painting continued, however, and some wonderful portrayals of court gatherings were produced by a group of painters, which included Muhammed Nadir, Goverdhan, Hunnar, Bachittar and Chittarman. There is a new sumptuousness in these works – a splendid example is *Shah Jehan Receiving the Persian Ambassador*. The draughtsmanship is meticulous; there is a frequent use of gold and the portraits are realistic.

Shah Jehan was ousted by his third son, Aurangzeb in 1658 AD. One of the pretexts the latter gave for his rebellion was that his father had beggared the Treasury for the sake of useless buildings. He banned all art in the name of Islam, and the court painters removed themselves to other capitals, or worked surreptitiously for various noblemen, concentrating on erotic art. Later in life, worn out by twenty years of campaigning in the Deccan, Aurangzeb relaxed his ban on art and even allowed a few portraits of himself to be painted. The *Siege of Golkonda*, actually painted during the action, exhibits the vitality and skill in composition of the earlier Mughal masters.

From the time of Akbar onwards, vassal princes from the Punjab and Rajasthan, from Central India and the Deccan, had visited the central court, first in Agra and later Delhi, to pay tribute to the emperor. They participated in court festivities and returned to their native states, bearing representations of the Mughals which had been presented to them as gifts. The Emperor Akbar had been a generous and liberal overlord, and the atmosphere of the native courts was much the same as it had been when the states were independent. This ensured continuity in indigenous religious practice, as well as continuity in music, painting and sculpture. No doubt the artistic influence of the capital affected the outlook of the native craftsmen, particularly demonstrated in the sophisticated miniatures and the minor decorative arts of costume design, armour and jewellery. The mural tradition was particularly strong in these states, especially the rendering of Hindu Vishnavaite, Sivaite and Shakti themes, and was continued, along with folk art decorations on walls and doorsteps of houses, and on palm leaves and cloth. Mughal influence acted as a catalyst in providing more direct court patronage. Portraits of princes were painted with Mughal-style profiles, and the epic *Ramayana*, *Bhagvad Purana* (Life of Krishna) and erotic literature, usually based on the Krishna cult, began to be illustrated. In Basholi, in the Punjab Himalayas, Mughal influence resulted in a highly distinctive style of Pahari (hill) painting, with burning red backgrounds (reminiscent of the 15th century paintings in Mewar), Krishna-blue, ochre-yellow and black. The impact of these Basholi paintings was felt in all the Punjab Himalayan states. Certain hill painters like Nainsunch in Jummu state and his brother Manax in Guber state produced under the Mughal impact some of the most sophisticated and intense paintings in Pahari art. As we have already seen, there was a primitivist tradition of painting in

Rajasthan, which had emerged from the folk tradition influenced by the priestly style of Jainism. In the 16th century, Mughal brushwork had an influence in Mewar and other native states of Rajasthan similar to its effect on the art of the Pahari area of the Punjab Himalayas. The miniature form was used to illustrate themes from the old epics and the Krishna legend, but the dense, highly emotional colours of the Hindu tradition gave them depth. The profile portrait was used to glorify princes and noblemen. The varying styles of Bikaner, Jodhpur, Kishangargh, Jaipur, Bundi, Kotah Sirohi and Udaipur all retain, in common with each other, the intensity of the devotional Hindu Bhakti tradition of worship of one god, but were enhanced by the Mughal finish. The synthesis between the Indian and Persian styles which Akbar had sought to achieve was realised in a number of masterpieces of Rajasthani painting.

Under the real impact of Western civilisation in the 18th century, the native two-dimensional religious tradition yielded finally to the three-dimensional style of portraiture, and portraits of individuals superseded the epic tradition. A hybrid Indo-European bazaar art developed, ending the poetic devotional attitude which had inspired the plastic arts; a manifestation of the changing life patterns of the country.

The history of the subsequent synthesis between the Indian sensibility and Western techniques and materials is the montage drama of the art of our own time.

Indian Architecture

E. Maxwell Fry

The basic culture of India, which has continued to exercise its massive pressure under whatsoever overlordship, is the product of geography and climate. 'India has', says Benjamin Rowland, 'an impregnable geographic isolation: it is in the shape of a great sealed funnel depending from the heartland of Asia . . . made for an inevitable retention and absorption of all the social and cultural elements poured into it.'

A people dominated by the dramatic effects of climate worship power where they find it, through divinities created out of the fabric of their life and watching over every part of it. Hinduism, assuming its dominant Aryan composition a thousand or more years before the coming of Christ, revealed an obsession for speculation on the meaning of life that led to the idea of the soul's absorption into the Universal Being, and by contrast, to the comparative worthlessness of the realities of day-to-day life. It is a pantheism of negation that incorporates the role of suffering, to which the Buddha added that of asceticism without disturbing the pantheism that is so much the necessity of Indian conditions.

To appreciate Indian art we must put aside our preference for the clear-cut outlines of Greece and Rome and enter a world of dream and suggestion stemming from the inner life of a mystic society, reading the figures and symbols as the pictorial script of India's ultimately changeless wisdom. And this art, exactly reflecting the society it serves, is as traditional as the division of social energies known as the caste system; its symbols may be re-interpreted but they stand immovable as the basic language of an art that is expository before it is aesthetic.

The first thing that we, as strangers to Indian architecture, notice is that it is external. Unlike Gothic and Classical church

145

architecture it is, with only the partial exception of the cave temples, an architecture made for external effect, with a general impression of so many overwrought tombs standing on the plains.

The *stupa* or relic mound, with its decorative gateway and railing, is of ancient origin, incorporating within its symbolism all sorts of facts and fancies out of the history of the village, through which an agrarian society expresses itself. The great *stupa* at Sanchi, with its dramatic contrast between round and square forms, is at once a tomb and a representation of the cosmos. There is in this smooth-swelling mount, culminating in a heavenly mansion, the loftiest expression of the ultimate; but its great barred gateway is alive with much more than the deep-cut carvings of the life of Buddha, for it teems with life itself, with the elephants, lions, peacocks, the fruit and foliage, and not least, with the loveliest representation of woman in the form of *yakshis*, or goddesses of fertility.

In these goddesses from the Sanchi gateway float human beauty that will recur in the Ajanta caves a thousand years later. Greece could do no better and its sculpture lacks one potent characteristic of the Indian delineation of flesh, the elimination of muscle in favour of a sort of tension that gives breath and life, accentuated by the sharp carving of ornament against the smoothness of skin.

The Mauryan period three centuries before Christ gave not only the column with its lotus head, its wheel and lion that the kingly proselyte Asoka so prodigally bestowed, but the *chaitya* or shrine with its curiously weak and sometimes ungainly arched opening that we may appreciate only when we know how closely it reproduces the village roof with the beams projecting and the thatch swept inwards to ward off sun and torrential rain. Here, as in Greece, ancient and homely constructions are monumentalised in stone, though in India retaining much more of the exactness of the original model. The deep carvings of the Sanchi gateway are echoed six hundred years later in the great *stupa* of Amaravati, sole remnant of a distinguished era with the same deeply-cut carvings, the figures layered one behind the other as though they occupied the architectural stage of a Palladian theatre, the naturalistic bodies swirling in dramatic animation.

It is difficult in so short an account to convey any true sense of either time or place, and I pass to the full flowering of the Gupta period with little mention of the classically inbued Gandhara style, feeling that, as in the earlier Maurya period, 'the political unity of India made for an artistic unity transcending regional boundaries'.

The *chaitya* hall now comes into its own, not only as a freestanding shrine but realised in the solid stone of a rocky gorge cut into its side, or within it as a man-made cave. The crescent-shaped valley of Ajanta contains many caves, built at different times for different sects. They are not so much caves as internalised temples that, cut downward from the living rock, are yet structures with all the appurtenances of support; and they are built up of the unchanging symbols from the *chaitya* itself, to the columns, the lotus heads, wheels, the countless Buddhas with their followers, or the whole panoply of Hinduism determined in its order by a rigorous and solemn architecture that is the match of anything anywhere in the world.

At Ellora is the rock-cut temple, one of the greatest monuments of Dravidian art. It is, though as wide and one-and-a-half times as high as the Parthenon, a monobloc cut from the rock face and standing in the pit of its own making. It is the monument of a mountain, the replica of Mount Kailasa in the Himalayas, Siva's external home, for it belongs to the final period of the Hindu dynasties that have submerged Buddha in the valhalla of Hindu mythology. From now on we may think only of the single guiding faith realising itself in countless temples in all parts of the subcontinent. They are of several types: the first and earliest is distinguished by its spire or *sikhara*, is conical and convex, and is crowned by a vase-shaped funnel or *kalosa* of which the temple of Kandariya Mahadeo, Khajuraho, serves as an example from among thousands. The second ascends in a series of horizontal terraces of which only the crowning *memba* is repeated on the corners of successive terraces; each, devoted to a separate deity, is a sort of miniature *stupa*. The Virupaksha Temple in the deserted city of Pattadakal is a splendid example, as is the unfinished temple at Konarak, the Black Pagoda, notorious for its erotic sculpture despite its semblance to the chariot of the gods. High up in the structure are monumental statues of

147

female musicians that are the *yakshis* of the Sanchi shrine, made in the vernacular of late Hindu art, achieved as a regulated whole before the inevitable baroque decline. The third type, which is restricted to the Deccan and the west, is the survival of the old Buddhist *chaitya* hall with its barrel vault. I close this brief description by reverting in time to the rock-cut cave temple of Elephanta in which is to be found sculpture that, were it not for the bias against Indian art, would rank with or even above that of Praxiteles. This temple is the Parthenon of India and, in the image of Brahma and the colossal wall sculpture of the Trimurti if nothing else, is the quintessence of eastern mystic art, the maturest expression of aspiration to be found in the civilised world: 'One sculpture like the Trimurti of Elephanta and one painting like the Prince Siddhartha from Ajanta', said E.B. Hanell, 'suffice to show that Indian artists may stand among the greatest in the world.'

Though the eighth century saw the beginning of Hindu decadence, enough temples were built to make the reputation of any country before temple building ceased with the establishment of Mughal rule over large parts of the subcontinent. From then on Indian craftsmen contributed to what was essentially an alien style of building. The style was Persian, though the early conquerors were Turks. It was a style well established elsewhere, though the first building of note built under its regime, the congregational mosque known as the Quwwat al-Islam (the Might of Islam) at Lal-Kot outside Dehli, was Hindu in style and craftsmanship. The Qutb Minar, close by and of the same period, is the last and finest of a series of funerary towers that scatter the route through Afghanistan to Iran and Anatolia. Indian craftsmen were used, but the style is no longer Indian.

The unstable political situation in the thirteenth century militated against monumental building and it was not until the reign of the great Khilji monarch Alla-ud-din (1296–1316) that building recommenced. He enlarged the Quwwat al-Islam and, finding Lal-Kot too small for his taste, moved on to a new site at Siri, the second of the seven Dehli cities. In 1320 the Khilji rule gave way to the Tughluqs and the fourth city was founded at Tughluqabad, with cyclopean walls designed to resist renewed Mongol attacks. It was, according to

148

the Moroccan traveller Ibn Batuta, the largest city in India, though it was abandoned in favour of a site near the Jumna river, probably because it lacked a good water supply. Ibn Bututa described the harsh life of the times, the ceremonious splendour of the court, and the terror inspired by the rulers. Warrior rule still prevailed with very little relaxation and the surviving tomb, that of Mohammed Tughluq's father, bears witness to this. Mohammed was succeeded by Firuz, who repaired the Qutb Minar, built several mosques and abandoned the Lal-Kot area for Firuzabad, another site on the banks of the Jumna, the ruins of which convey very little of its original splendour.

Little building of note took place under the succeeding dynasties of the Mughals, Saiyyads and the Lodhis, although the Lodhi tombs in the Lodhi Gardens in Dehli are finely monumental and well-proportioned and there, too is the octagonal tomb of Sultan Sikander within its stongly-walled garden. The great period of building, however, followed the establishment of the Mughal Empire that endured from the early 16th century until well into the 19th century.

The Mughals were not necessarily mongols. They shared in the brilliant life of central Asia, a culture of war and art centred upon Herat, from whence came Babur the first Mughal conqueror and his son Humayun, who ruled after him before submitting to the Afghan Sher Shah. The latter was one of the ablest rulers in India and his administrative structure lasted four centuries. His impressive mosque remains within the crumbling walls of the Purana Qila, a great fort of which only the gateways remain. This and the Lodhi tombs are what remain of Afghan rule.

Humayun was cast out of his capital and in the most desperate period of his life his wife Hamida Begum bore him a son, Akbar, who lived to be first the prince of a restored Empire and then the greatest of the Mughal conquerors. His father's tomb is a landmark in south-east Dehli, the earliest of a succession of great garden tombs that in size, style and treatment set them apart from the past and inaugurated a major development of Indo-Islamic architecture. The tomb of Mohammed Shah in the Lodhi gardens, with its prominent *chujjas* or eaves and the ring of domed pavilions reminiscent

149

of the *chaityas* of Hindu temples, has Indian elements, but Humayun's tomb is pure Persian, from its gleaming white dome to its tile-encrusted colonnades of ogee arches; a Persian building in a Persian garden.

Visitors coming to India for the first time discover in their inevitable excursions to Delhi, Agra and Fatehpur Sikri a collection of architectural marvels that, including as they do the forts at Delhi and Agra, the Imperial Palace at Fatehpur Sikri, the tomb gardens of Akbar and the Taj Mahal, present so strong an impression of an overwhelmingly successful culture as to obliterate all else. What strikes them is the comparison between the grandeur of the outer walls, the awe-inspiring symmetry of the ceremonial approaches, and the tender delicacy of the palace interiors in which red sandstone gives way first to gateways and facades of ceramic tiles where a pattern of never-ending intricacy is controlled by huge scoops of shadow and geometric outline, and finally to white marble, fretted in pierced screens or inlaid with floral design of unexpected naturalism.

This is the impression given by the palace forts of Delhi and Agra, but at Fatehpur Sikri, Akbar, made secure by conquest and wise rule, built a palace combining Indian and Islamic architecture which, being vacated no sooner than built, is exposed for us in all the details of courtly living. Its modifications of style, in favour of climate no less than Hindu tradition, show the workings of a great mind open to impressions coming to it from more directions than one. His own tomb at Sikandra nearby is the culmination of a cultural synthesis supported by poetry, painting and a whole range of lesser arts stimulated by a man searching for a new basis from an alien rule of rigorous racial and religious conventions. He was succeeded by rulers – conventional and less able – who nonetheless continued the tradition and added still further to the glory of Indo-Islamic art and architecture. The Taj Mahal may stand witness, as it did for the aged Shah Jehan, to the continuing vitality of the dynasty; but it is no fusion of styles, being pure Persian raised to the highest levels of taste by the use of Makrana marble and only differentiated from its gaudier counterparts in Iran by small touches of Hinduism, and by its dreamy setting on the banks of the Jumna river. Over-familiar as it is, its four lighthouse-looking minarets, hard enough in stark sunlight, the combination of the strength of the wide podium carrying the square form of the building and the immaculate proportions of the Tatai dome sailing over it, disarm all criticism. Strength and delicacy are once more combined in an enduring unity, and mere size, so often the means of domination, enclosed in a formal garden extending, defining, limiting and concentrating the effect of the central casket deep within which lies the tomb of the beloved wife, is transmuted into pure harmony. It is as though at last the practical and finite world of Islam were touched with the mystery of ancient India.

All over the north of India stand the remnants of a rule of which the final phases were supervised by the armed traders of Europe consolidating into the British Raj.

The architecture of this new rule as recorded by Sten Nilsson in a sumptuous volume, *European Architecture in India* (Faber, 1968), is by comparison with what went before a mere episode, an architecture transposed from its European setting, a second transposition from its origins in the Mediterranean in fact, to the unwelcoming banks of the Hooghly river or the burning plains of Bengal. It was by definition an effete architecture unsupported by faith and with no succession. Its prototypes in Europe representing a valid culture, were built in enduring materials and survive the passage of time, but these vainglorious reproductions, though they made in their time and place a brave show and were neither mean nor conciliatory, were alien importations tricked out with brick and plaster to comfort sick alien hearts, and decayed from the moment of their inception. God knows I feel their sweetness, coming upon among the fungus-rotted decrepitude of Calcutta, for they look as I feel, and as so many an Englishman has felt whose life was spent among those for whom the classical culture of Europe meant little or nothing at all. There are not enough of them left to give a coherent picture of early European rule and one must rely upon drawings such as that of the sea front at Madras by John Gantz or the lovely views of Esplanade Row and the Chowringhee, executed by Thomas and William Daniell.

These and Government House, Calcutta, reproducing

Kedelstone House; Serampore College, barracks at Dum Dum and elsewhere, and countless rest houses, circuit houses and the small paraphernalia of a widespread administration point to an acceptable level of classical taste deployed for the benefit of a rule of law, order and a conception of decency that has outlasted them all. As the nineteenth century progressed the architecture of the governing power disintegrated up to the moment of the final demonstration of grand scale imperialism when Sir Edwin Lutyens, in uneasy collaboration with Sir Herbert Baker, designed the vast layout of New Delhi centred upon a group of government buildings of real distinction. Using the red sandstone that lies at hand for all large-scale works, Lutyens, who was an artist capable of being obsessed with an intricate geometry of proportions, evolved an amalgam of classical and Indo-Islamic elements in which the *chaitya* interrupts the horizontal skyline in pyramidal forms, and the *chujja* scores deep shadow lines across his cliff-like façades.

Lutyens was responsible for the colossal overall layout with its axis centred upon the Viceroy's House, flanked by Baker's secretarial buildings, and this equal partnership of responsibility led to the visual eclipse of the Viceroy's House on the ramp upwards to the plateau of Resina Hill upon which all three buildings were accommodated. No amount of argument or pleading could overcome this disability once incorporated into the design, and only the grand imperial dome appears above the asphalted ramp between the secretariats where the whole building should have crowned a grand tripartite display.

Even so the quality of Lutyens' building is not in doubt. It is indeed in the grand English tradition so touched by Indo-Islamic art as to make it an acceptable inheritance of which the government of independent India might well feel proud, so honourably was it come by. The imperial gesture of New Delhi was indeed short lived, and in 1947 Gandhi's campaign of civil disobedience culminated in the bloodless transference of power, followed by the bloody detachment of Pakistan from India. The state of Punjab was cut in two with its ancient capital Lahore on the Pakistan side of the border. It was the rump of a refugee government huddled into the unsuitable

152

Detail: Radha and Krishna in a bower. Early 18th century Mewar school Rajasthan.

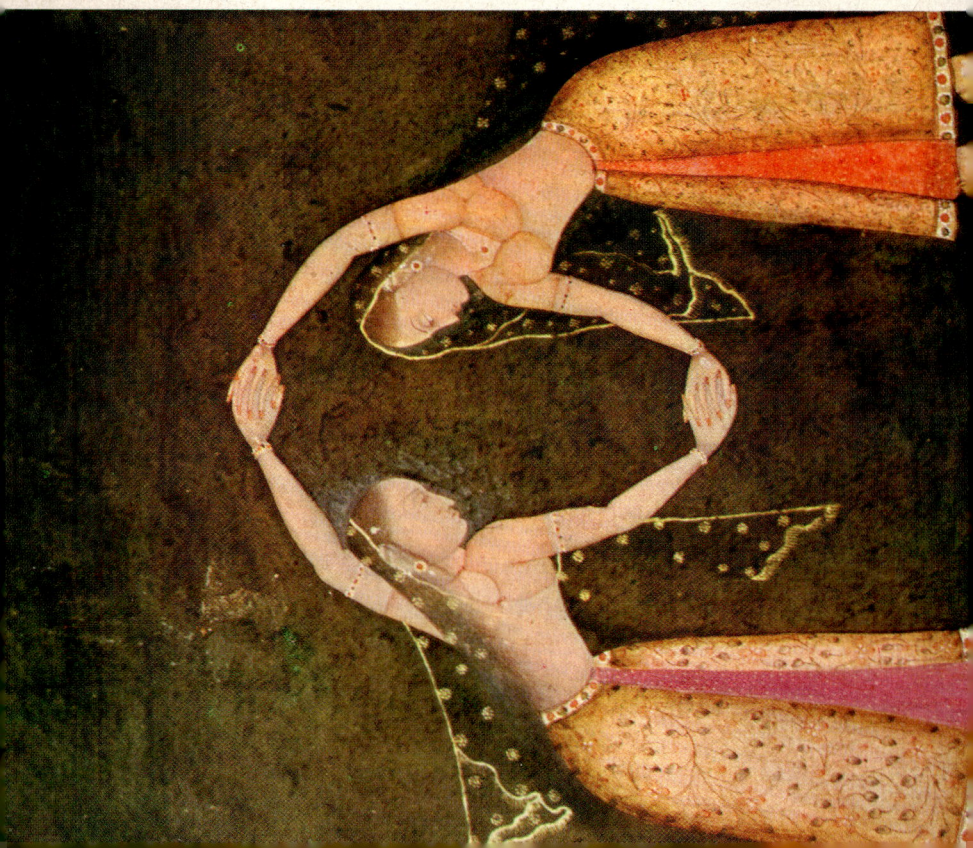

INDIAN ARCHITECTURE

summer retreat of Simla that decided to build a modern
capital at the foot of the Himalayas to be called Chandigarh,
and invited us with the great French-Swiss architect le Corbusier to do it.

Chandigarh is a modern city in an ancient setting, without
any of the real pressures of industrialism or motor traffic,
and with no skyscrapers. Built by hand with a labour force
that Akbar might have recognised, it is still a modern city
with electricity and sanitation and a geometric road system
enclosing wide sectors of residential land sealed off from
rapid and dangerous movement and locally self-contained.
At the head of the plan and against the background of the
Himalayas, Corbusier designed the famous group of capitol
buildings of which the High Court with its floating roof and
deep shadows is most closely assimilated to the conditions of
climate and the spirit of the country. The example of his works,
both here and at Ahmedabad where he was welcomed, combined with the deep experiences of the associated staff of young
Indian architects working on a new city of 150,000 people,
has had a profound effect upon contemporary Indian architecture. It has freed it from the old engineer domination and
made it an instrument for the major works of government,
including the new centres consequent upon a policy of industrialisation, and the future is with it.

Spread out over this great subcontinent lie the ruins of
civilisations that reach back to periods of western pre-history.
Dynasties have risen and fallen, sometimes to heights of
clarification, as in the caves of Ajanta and Elephanta, never
excelled by man. But buried deep within the manifestations of
art and architecture lies the molecular structure of the Indian
village, persisting under whatsoever form of coercion or op-
pression, achieving beauty in submission to the harsh laws of
nature, continuing as the humblest but the strongest basis of
an indestructable order upon which great edifices may rise.

*

153

*tails from two early 18th century paintings. Left: Women dancing. Provin-
*l Mughal from Rajasthan. Right: Women with wands of flowers: an illustra-
*n to the musical mode, Gauri Ragini. From the Deccan.

Indian Costume

Tara Ali Baig

For five thousand years of recorded history India has been a sort of inverse cornucopia, with people from other lands flowing into her, staying and becoming Indians. 'Know that knowledge whereby one sees in all beings immutable entity... a unity in diversity' was written in the *Bhagavad Gita* some three thousand years ago, after the Aryan invasion, but the writer somehow anticipates the coming of Alexander and the Greeks, the Huns, Arabs, Turks, Mughals and the 16th century freebooters and traders from Europe.

We have some idea of what these early peoples wore from ancient sculptures, the frescoes of Ajanta and Bagh, and the literature of Vedic times. Since clothing was associated with man only, the Vedic Indian sensibly considered that clothes must be connected with divine mysteries. Having imported male dominance into India's then largely matriarchal Dravidian culture, the Aryan further assumed that male garments were presided over by the deities. Different parts of the garment were therefore consecrated to different gods, while the borders and middle portion were considered the preserve of the ancestors. The fashion of wearing borderless *saris*, favoured by young glamour girls of Bengal in the thirties, evoked great wrath, but few realised that their horror was stimulated by a primitive memory of insult to the ancestors. Religious custom everywhere has attached great significance to clothing as a part of ritual, as in the Catholic Church. The ancient Indian was no different, he could make no sacrifices without the appropriate garments, but went a step further and introduced ritual clothing as part of his daily life.

In ancient times, cloth draped around the hips and a shoulder scarf were common to both men and women. This cloth could be doubled back to make a mini-*lungi*, worn in

154

Kerala to this day, or could make a maxi with full pleats as worn by the modern Bengali male. The pleats played a very significant part, the gathers were considered an obstacle to evil; the folds of the scarf covered the heart, the folds of the turban covered the head, and the folds of the *dhoti* or *sari* covered the reproductive organs – thus all the vital centres were protected. But although women decked themselves out in enormous earrings, ropes, chokers and girdles of gold and gems and wore turbans and high head-dresses, they generally wore nothing more than a cunningly-draped lower garment which swathed the hips tightly and was drawn into a jewel-belted panel in front. The epic *Mahabharata* tells us that Queen Draupadi could not appear before her elders, not because she was bare-bosomed, but because her *dhoti,* wrapped twice about her thighs, indicated that she was in her periodic condition of uncleanness. Though the outer garments have now changed considerably and the bodice or blouse is almost imperative for peasant women, the undergarment remains more or less unchanged from the time of the *Mahabharata.*

Perhaps the most interesting feature of the use of the excess cloth of the female *dhoti* as an upper body covering, is that it foreshadows the *sari.* The first clear proof of this is a sculpture in the Amaravati Buddhist stupa (2nd century AD). In Nagarjunakonda, there is a stone panel showing unmistakably *sari*-like garments. In Buddhist literature three items of clothing were listed: the *dhoti,* the head-dress, and upper cloth, and when it was cold, a long-sleeved, flare-skirted tunic. Mention is also made of the undergarment in these texts; a form of *langot* for men and women which has not much changed.

Two popular myths about Indian dress relate to the *sari,* and to clothing made with the help of a needle. The former was widely believed to have been introduced by Alexander the Great, but to attribute the permeation of Greek style into India's closed and stylised circles via a conqueror without womenfolk, with only a governor or two to guard his conquests in India, is asking too much of history. The needle was simply an invention of cold climates. South India to this day could practically dispense with it were it not for a much more pernicious influence on the subcontinent: Victorian prudery.

155

As late as 1881, Bengali women who followed the new reform religion of the Brahmo Samaj created a great scandal by wearing blouses. In select circles, covering the upper body with anything but the sari cloth was considered a prerogative of prostitutes. That the blouse is now worn in every part of India by tribal people and in villages and cities, shows that the last century has seen a radical change in the attitude towards clothing.

In 300 BC Megasthenes wrote of Indian love of ornamentation and the brightly dyed muslin garments worked in gold and precious stones. Men and women wore jewels in great profusion, from rows and rows of long heavy necklaces and earrings, which show in sculpture after sculpture the redundancy of the blouse, to wide jewelled belts, hip chains and ornate head-dresses.

Kalidasa's statement that 'garments for women were intended for appreciation by their beloved', may explain why some sculptures show women with only a hint of covering over the legs, which are outlined in jewels, while others show no attempt to cover the body at all. One can only assume that peoples from Kalidasa's times (3rd century AD) to that of Mughal beauties who posed seductively in the most diaphanous garb, though covered with exquisite jewels, regarded the female body perhaps as the Greeks regarded the male; a matter of art and perfection, to be glorified in painting and sculpture.

In one ancient text the judges of two dancing masters decreed that the girls must wear very thin garments so that the grace of their bodies might be displayed. There is also a vivid description of the old city of Dwarka where the visitor describes women wearing blouses delightfully revealing their charms, and perhaps the oldest description of a woman's costume in detail comes from Banabhatta, who writes of Savitri standing by the side of Brahma, wearing 'a very white and fine garment made from the tree of paradise and she had put on a piece of cloth made from lotus filaments as her upper garments.'

But seductiveness in those days seems to have been concentrated far more upon the navel. A 14th century commentary states: 'If a lady looks longingly at her navel, or if the knots of the pleats of her lower garment gets loosened or if she frequently unties or does up her hair when she sees her lover's friends, she indicates her passion for her lover'......(all this in sanskrit with words so long (the first in the sentence: *nabhimulakucodaraprakatanavyajena*) that they must surely dampen ardour. The cloak described in these old books was apparently some kind of tailored garment covering the whole body, something like the *chador* of modern Iran. It seems unmarried girls wore it when going out and its ancient name was *coloka* which is, interestingly enough, very similar to the word *cloak*.

If the draped lower garment, flimsy breast band and shoulder scarf constituted the basic ingredients for clothing in ancient times, from the Gupta Dynasty around 320 AD onwards, new influences infiltrated with new peoples. Gupta coins show men wearing clinging trousers, in northern India today called a *churidar*. The *dhoti* and baggy or tight trousers continued side by side, but the trouser took about one thousand years to be accepted as a valid and distinctive regional style of dress.

What is sometimes overlooked in learned treatises on the influences brought into India by conquerors or marauders is the simple element of trade. Egyptian tombs contain ample evidence of Indian muslins, wool and silk fabrics, but perhaps more significant is the amount of travel and contact in ancient times between the peoples of India, West Asia, Greece and Egypt. Any close look at the clothing worn in these countries in ancient times will reveal the mini *lungi* and lack of upper covering in Egyptian frescoes, the cloak and tunic of Gupta India in sculpture of the Euphrates Valley and the draped Roman *toga* or Greek dress worn by women and men, so similar to the draping of the *sari*. The women of nomadic Iranian tribes wear full skirts even today with ribbons and borders similar to the skirts of Indian women in Rajput and Pahari miniatures. The shoulder scarf, cloth or veil seems to have been common in one form or another to all these cultures, and the people of Kalidasa's time sometimes exchanged their shoulder garments as an act of friendship.

Along with the trouser in northern India came variations of the tunic. In the 6th and 7th century temple sculptures of

Mahabalipuram near Madras in the close-fitting tunic with tapered sleeves, also seen in the earlier Ajanta frescoes, there is evidence of the future *angarkha*, the fashionable Mughal dress considered with its thin fabric, side fastenings of a special yoke, and full, calf-length skirt. *Angarkha* means protector of the body and the names of all other garments signified their use. The *cummerbund* is simply a band to bind the waist or *cummer*. And the modern *bundgala* the high necked, evening coat is what it says: closed at the neck. The word *bandanna* comes directly from the Rajasthan tie and dye technique of dyeing fabrics known as *bandna* or tying – another proof that traders cross-pollinate cultures more effectively than conquerors.

By the end of the Mughal period, which by the 17th century had imported Persian, Arabic and Turkish ways of life, clothing had undergone considerable changes in the north, in Bengal and on the Madras coast. Added to draped garments, short, long, full or scanty, from the male *dhotis* and *lungis* to the short *saris* of tribal women in Orissa, the tucked-up ones of coastal people, and the more usual six-yard type, were added the *pyjama* and *kurta* (a long shirt), the *achgan* (a long high-collared coat), the *choga* (a full-skirted coat still worn in *Kashmir*), the full, long skirts of women in Rajasthan, and all the elegance of the Islamic courts. Between 1500 and 1700 AD the idle life of rich women in *seraglios* led to all sorts of dizzy styles. There were gauzy over-dresses tied with tassels on the sides; tightly fitting brocade trousers for slender legs, short and cunningly fitted blouses that displayed curves and stomachs, and narrow tight trousers with overlays of fine muslin skirt. A story is told of the Princess Roshanara being chided by her father for being indecently dressed. She retorted that she wore thirteen layers of Dacca muslin, but this fabric was so gossamer-fine that yards could be pulled through a wedding ring. In most Mughal portraits the *ordni*, veil, is embroidered and edged with gold threadwork and carried over the head. This was nothing more than a gorgeous replica of the shoulder scarves and veils worn from ancient times in India. While in aristocratic families it played the role of revealing more provocatively than it concealed, among the working classes it protected the eyes from dust, the head from the sun

and the baby carried on the hip. Later, as a result of the re-
pressions and restrictions of society, the veil covered the face
entirely. Today the veil is synonymous with the seclusion of
women, its glamour and utility forgotten.

Draped clothing, trousers, baggy and tight, scarves, small
blouses, shirts, full skirts and *lungis* made up the main ingredients of clothing on the Indian subcontinent. In addition
to ethnic divisions, dress began to be defined also by linguistic
areas. To many Hindus, the Mughal-style clothes worn by
Indians in the north were not Indian at all; Islamic costumes
were considered foreign. This prejudice does not bear close
scrutiny, for many garments thought to be Islamic were used
in ancient India.

Whatever the local resistance to new ways of dress, history
ruled taste like a dictator. The Mughal rule began to crumble
and the Marathas, under Shivaji, fought to regain Hindu rule.
The Marathas had always been hardy and warlike and their
women rode horses which accounts for the special way they
tied their *saris*. Unlike the *sari* style elsewhere in India, except
the coastal regions or where women worked in paddy fields,
the Mahratta *sari* was worn like the male *dhoti*, the folds drawn
between the legs and tucked tightly behind.

By the time the British consolidated their rule in India, the
dress of the people had settled into fairly fixed regional
patterns. Where Mughal influence had been strong, men wore
baggy or tight cloth trousers with an overshirt and long coat
and on their heads high turbans with a long streamer behind,
or conical caps called *kullah* or the small boat-shaped one
similar to the later Gandhi cap. Women wore the courtly tight
pyjamas with tight *choli*, a sort of breast-band with sleeves and
a gauzy overshirt and veil.

Where Hindu customs remained strong, the *dhoti* was the
recognised dress of men and the *sari* in many variations was
worn by women: the nine-yard variety in the region around
Bombay and some parts of central India and the south; the
Bengali style which draped the end piece over the right
shoulder to the front instead of the usual way; the short
Orissa *sari* just below the knees, and the Gujarat style adopted
later by the Parsis. Blouses were not originally worn with the
sari but all types evolved in different regions. The Saurashtra

159

style is probably the most attractive, fitted cunningly to a high bosom with tiny cups and insertions and tied with tassels across a bare back. The most original is undoubtedly the long-aproned blouse of the Lohars, a tribe of iron-worker nomads who left Rajasthan at the time of the defeat of Raja Mansingh declaring they would not return till he was king again. That was some four hundred years ago. They still go about the countryside in their splendid old carts adorned with brass decorations and huge wheels studded with the signs of the zodiac.

The most common type of blouse worn in different parts of the country, once the present method of draping the *sari* was widely accepted, is a tight-fitting bodice with sleeves, fastened down the front. All sorts of variations to this have been created by fashion-conscious women over the last fifty years, imitating western styles at the turn of the century. And now to the modern scanty *choli* they have added the *hipster sari*, drapped well below the navel, reminiscent of the statues of ancient Mathura.

British rule caused many radical changes in dress and the masculine *dhoti* was replaced in urban areas by shirts and trousers or the European suit. If the non-suited elements of society wore trousers with the shirt tails hanging outside, it was a form of male modesty. A race memory too perhaps, because for thousands of years, male dress in India always covered the wearer fully above the knees. Whereas the upper body could be bare even in public, it was unthinkable for a conservative person to wear his protective shirt tucked into the pants.

Women's dress, apart from the few aberrations of the blouse, was never affected by European influence. Young people took to sports-wear such as shorts and slacks, but every woman was instinctively feminine enough to realise that the *sari* was pre-eminently feminine and graceful. It is elegant in a variety of fabrics; silks, cottons, organza, net and nylons. The normal *sari* is 6 yards long and 45 inches wide, it can be printed, hand-painted, dyed in every jewel colour, embroidered with silks and gold thread, woven in every conceivable design with borders woven in or sewn on and it looks elegant with anything – except the twin-set. Although South India's climate is always warm, some of the heaviest, most beautiful silk *saris* are woven there and the characteristic *conjeevaram* silks are popular all over the country.

Brocades associated with the great weaving centre at Benares have now lost some of their importance in Indian dress with the substitution of sober *bundgala* or *Jodhpuri* coats for evening wear. The *bundgala*, sometimes erroneously called the Nehru coat, stems from the *achgan*, for it has the same high fitted collar and buttoned front, but is as short as the European evening dress coat. Benares workers continue to make magnificent gold and silk fabrics for evening coats, stoles and *saris*, especially the cloth-of-gold wedding *saris*, for no Indian bride of any class can face her bridegroom unless she is suitably adorned.

The distinctive signs of a married woman vary from one part of India to another. The married Gujarati women for some reason wore an embroidered panel down the front of the *sari* as part of the pleats. In old Maharashtra, a high-born married woman would drape her *sari* over her bosom, and Bengalis would distinguish the married woman by a line of red powder above the ceremonial red beauty spot on the forehead, taken right through the centre parting of her well-oiled hair. Red is the wedding colour in India and white the colour of mourning. Black was always considered very inauspicious; only tiny babies might be found swathed in hideous black bonnets of cheap satin with tinsel to ward off the evil eye. And the black collyrium, made from candle flame on a copper vessel rubbed with garlic, rimming the eyes of children, is said to give strength to the eyes and deflect the main force of evil from the child's face. Certainly those dark rims on baby faces are startling. Their older sisters who use eye makeup so effectively (the principal makeup used by Indian women), are clearly unconcerned with the evil eye.

Using all the costumes of the past, the young Indian girl today creates many new styles. Girls in the main cities are especially fashion-conscious and young designers are springing up who hold fashion shows modelled with great panache by girls with no professional training whatsoever. While the flood of western hippies into India has shown how badly Indian dress can be worn – for hippies inevitably 'go native'

161

and slouch about in sandals, *lungis* and loose hand-spun shirts – the smart set has taken to transforming old *sari* panels, brocades, prints and fabrics into long *lungis* worn with soft silk shirts. The loose white cotton pyjama with *kurta* (shirt) and shoulder scarf has now given way to pant suits based on Mughal miniatures where the tight trousers (cut on the cross) match the veil, and the elegant overshirt is beautifully fitted. Boys have also taken to brighter clothing, and the bush shirt, imported from the African safari dress, is admirably adapted to Indian climate. The need for a shirt outside the trousers, has gradually given way to the Lucknow embroidered muslin shirt, or the high-necked hand-spun coloured shirt or silk shirt with the Nehru waistcoat. The latter is equally at home over trousers or the flowing Bengali *dhoti*.

The young have not yet fully explored the bewildering wealth of tribal costume. The 1972 census figure for the tribal population of India remains to be seen, but one census listed them at 25 million. These forest tribes have their own ways of life and dress. In the North-Eastern Frontier Agency and Naga region they wear black, white and red short, woven *lungis* and high head-dresses with feathers; and in Manipur the women wear the most exotic costumes. Other regions have the short *sari*, often minus the blouse, and among the Santhals, an elegant draping of the *sari* revealing one shoulder and breast. Tribal costume is often highly ornamented with exquisite earrings made from golden grass, as among the Ahirs of Central India, or shell and flower jewellery and heavy gold and silver ornaments in extraordinarily modern design. There are hill tribes in the Himalayas who wear woollen coats with flared skirts tied about the waist with ropes of thin wool, fifteen yards long. These people wear round caps with a fold-back rim and the women wear long wool and heavy cotton robes and massive jewellery. Dress, like names, continues to be an accurate index of where people originate. Names give you caste and region, and dress even indicates religion. So distinctive has this become that an observer claims he can tell an Indian Christian woman by her hair style.

The *Bhagavad Gita* spoke of 'unity in diversity'. In a loose sense, the unity does cover India's vast storehouse of diversity, for India's cornucopia is a little microcosm of the world. In terms of dress, the synthesis of so many ancient styles now achieved by the selective choice of the young has merely anticipated what is happening elsewhere in the world. More and more people insist on dressing as they please, not as fashion dictates. The battle of the mini and the midi in Europe has led to a do-it-yourself and please-yourself attitude to clothes. Indian conservatism has maintained regional dress, while the individualists, from the holy man to the eccentric, dress exactly as they wish. Clothing today is both covering and adornment.

If the urban clerk, small-time factory worker or middle-class man goes about in shabby semi-European dress, often drab, women somehow manage to invest even poverty with gay colours and inherited finery. There is nothing gayer than a group of Rajasthani building workers, their children perched on their shoulders, brass lunch vessels in hand, heads covered with variegated veils, swinging along, their full skirts aswish and silver anklets gleaming in the sun. Women's Liberation may not have reached them yet but they are the embodiment of *sakti,* the ancient attribute of women . . . energy and strength.

The Classical Dance of India

Indrani

The classical dance of India is a unique and distinctive twofold art of pure dance and rhythmic acting. Pure dance is composed of rhythmic dance movements that have no symbolism or meaning. Rhythmic acting is composed of dance that tells a story or interprets a poem, with facial expressions, symbolic hand gestures, and body movements, combined to form dramatic dance. The gesture language and mime of the dramatic dance is a mirror of the life, moods, and passions of humans and of the gods, who are conceived in the human image.

In tracing the history of dance in India, and the origins of the classical dance in particular, we find that a simple form of music (the instruments were flute, drums and lute) and dance prevailed in the early pre-Aryan civilisations such as Mohenjodaro. A witness to the dance is the famous copper figurine of the dancing girl of Mohenjodaro (2500-1500 BC)

Whether the Aryans brought a form of dance with them when they infiltrated into India, or whether they absorbed the dance of the local people whom they conquered, we can only speculate upon. However, the first of the four *Veda's* (the sacred books of the Aryans), the *Rig Veda* of 1500 BC, is filled with references to dance, describing gods, goddesses and celestial beings as dancing excellently and revealing that during marriages, festivals, and sacrifices, dance played an essential part. Professional dancers, both men and women, danced solo and in groups, and were an accepted part of Vedic society.

In the later Vedic period 1200 to 800 BC, the defining of the Margi and Desi types of dance begin what was to be a clear distinction between classical dance and popular dance. During this time religious ritual before the Vedic altar became

164

increasingly complex and involved, using symbolic gestures of the hands to accompany invocatory ritual chants; this was undoubtedly the origin of the unique gesture language of the classical dance which was to follow.

In the period of the great epics, the *Ramayana* and the *Mahabharata,* 600 BC to 400 AD, the classical dance is established as a stylised, systematised fine art, performed in *sabhas* (assembly halls) and patronised by royalty and aristocracy.

The earliest sculptural representations of classical dance in existence are from the 1st century AD in the palace and theatre rock caves of Udayagiri where the stone frieze depicts a dancer in classical dance pose and gesture, accompanied by her musicians playing the flute, stringed harp and drums. Similar dancers are found in bas-relief in stone at Bharhut, with musicians playing the cymbals and the seven-stringed harp.

The *Natya Sastra* of the second century AD is the oldest available text on Natya, 'the dramatic science' compiled from ancient sources by Bharata. The art as defined by Bharata must have been practised for several centuries prior to its being actually codified in the *Natya Sastra*; Bharata himself quotes from many earlier texts, which are lost.

Bharata states that drama must include dramatic dance, pure dance, song and musical accompaniment. All the dance movements of the body are defined and classified in the *Natya Sastra* in precise choreographic terms.

The units of dance movement are called *karanas.* Six or more *karanas* make a dance sequence (*angahara*). There are 108 *karanas* and 32 *angaharas.* Sculptured figures of the *karanas* are carved on the inner walls of the Siva temple at Chidambaram, in South India. Some of them survive in the movements and postures of traditional contemporary dance.

Rhythmic gaits and steps (*charis*) include movements in space as well as floor contacts. Turns and pivots are called *bhramari* or 'bee movements'. There are 32 *charis.*

The *angas* are the movements of the body. There are 10 arm movements, 10 head movements, 5 of the chest, 5 of the waist, and movements of the thigh, calves, knees, wrists, and elbows are also defined. The movements of the separate parts of the body are identified with the sentiments and are useful in dramatic dance as well as pure dance.

165

Indian aesthetics define sentiment (*rasa*) as of 9 kinds and these are named by Bharata as: romantic, comic, pathetic, furious, heroic, terrible, odious, marvellous, and peaceful. These are called the permanent sentiments (*rasas*), the source of emotional states and transitory moods of drama and dramatic dance. Nowadays dancers master these sentiments in the form of 9 basic facial expressions from which innumerable subtle varieties of expression make it possible to interpret manifold moods and themes, depending on the dancer's acting ability.

Eight movements of the eyeballs are identified with the sentiments and these are the source of the glances that arise from transient moods of pleasure, anxiety, yearning, etc. The movements of eyebrows, eyelids, lips etc. in facial expressions are concisely defined in the *Natya Sastra*.

'Where the hand moves, there the eyes follow; where the eyes go, the mind follows; where the mind goes, the mood follows; and where the mood goes, there the flavour (sentiment) arises.'

The language of gestures makes use of single and combined hand poses in dramatic dance and drama. Bharata describes 24 gestures of the single hand and 13 combined-hand poses.

There are a total of 32 single-hand gestures and more than 24 combined-hand gestures, including those mentioned by other authorities. All of these gestures are in common use in dramatic dance and dance-drama. One might call the single-hand gestures an alphabet of the hands, and the combined-hand gestures a dictionary of the hands, for almost any object and any sentence can be interpreted with gestures of the hands, variously combined.

In the *Natya Sastra* it is stated that although the basic choreographic science is recorded in its text, yet the different parts of the country have different versions of the dance technique. Even today a variety of classical dance styles exists in India, handed down through the years by hereditary dance masters and dancers, with ever creative and changing traditions.

Some of the major classical dance styles of India in existence are the *Bharata Natyam* solo feminine dance of Tamilnad; the *Kathakali* performed by male dancers; the *Kathak* dance

166

of north India with a strong Muslim court influence; the *Manipuri* group dances of Manipur; the *Orissi* of Orissa, and the *Mohini Attam* of Kerala, both solo feminine dances.

Bharata Natyam, the classical dance of south-eastern India, is a solo style of dance and mime performed by women. Traditionally a temple art of ancient origin, it has been handed down for centuries by the *Devadasis* who, as votaries of the gods, dedicated their lives to temple service in South India. They not only performed the dance rituals in temples, but on special occasions also danced at the royal courts and received titles and gifts from the king. Their exquisite art has now re-emerged as *Bharata Natyam* on the modern stage and is being taught by Nattuvans, the hereditary dance masters who were formerly attached to temples and palaces.

Bharata Natyam is the most beautiful and captivating form of feminine dance in India. Its pure dance form expresses the infinite varieties of rhythm in beautiful plastic curves and flexions, precise rhythmic patterns of hands and arms, supported by complex footwork.

The basis of dance composition and its perfect exposition is the *adavu*. A rhythmic movement of the feet, a harmonious movement of the body combined with a patterned movement of the hands is called an *adavu,* indentified by rhythmic syllables (*ta tai tam-dhit tai tam*). There are 10 kinds of *adavus* in 12 varieties in each class – a total of 120 basic dance motifs.

The deep knee bend or 'pile' is the characteristic stance in rendering the *adavus.*

The dance always begins with a gracious erect pose, the rhythmic swaying movement of the neck, and delicate movements of eyes and eyebrows.

Rhythm patterns of the dance called *jati-adavus* are composed by the dance master or Nattuvanar. He conducts the dance performance, plays the cymbals, and recites the rhythmic syllables (*jatis*) that are played on the drum. The bells on the dancer's ankles articulate the rhythms of the feet. The entire choreography of each dance has been fixed by the dance master and nothing is left to chance or impulse.

In timing there are three kinds of tempi; slow, medium and fast. Slow tempo is doubled for medium, and quadrupled for quick timing. Within the timing cycle a four-beat rhythm may change to six beats, or any cross rhythm. All three timings, slow, medium and quick, are essential for contrast and rhythmic variety in the dance.

The dance repertoire was brought to its present form by four dance masters and musicians, brothers, more than a hundred years ago. The various dances were arranged for performance before the last Tanjore kings, who were great patrons of the art.

Gesture songs are always included in a *Bharata Natyam* performance. The songs may be purely devotional, addressed to a deity, or they may be love songs. The dancer must be highly skilled in the use of the language of gestures and the exposition of sentiments.

Hasta Prana, the 'Lives' of the hands, are those movements of the hands and fingers that give life and meaning to gestures. The cyclic turning of the hand with wrist as a pivot directs the palm up or down, forward, sideways, or toward the body. The fingers are disposed, bent or separated as required by the dramatic theme.

Gestures of the hands play upon the imagination of the spectators depicting moods, objects, and scenes, and the spiritual qualities of deities.

The *bhakti* cult – the path of devotion to God – brought forth religious songs in South India called *kritis* and *kirtanas* composed by saintly musicians.

The beautiful *kirtana, Natanam Adinar,* describes the wondrous dance of Lord Siva in the assembly of the gods. This song is meant to be danced as it contains groups of rhythm syllables interspersed in the text so that the dramatic dance may be combined with pure dance.

When the Krishna cult was propagated in south India in the 14th century the lyrical and impassioned songs of Krishna, the pastoral god, and Radha, his beloved, gave a different interpretation to the *bhakti* cult.

It is said that 'the love of man to God is as the love of a woman to her lover'. To adore God as the lover is the underlying thought of erotic poetry in India. In the dance the dancer places herself in the role of the heroine and addresses the hero both as beloved and God alternately for, according to Hindu

168

The river goddess Yamuna, one of the 8th-9th century sculptures in the Hindu temple in Cave 21 at Ellora. There are Buddhist, Hindu and Jain temples here 34 in all, carved into the slope of a hill.

philosophy, love in its highest form is godly.

Kama (desire) lies at the very root of creation. The goddess – consort of the deity – represents the manifestation of cosmic desire.

Kama is the Hindu God of love. He rides a parrot and bears a bow of sugarcane. His five flower-tipped arrows are the five senses.

The *Gita Govinda,* a beautiful Sanskrit poem of the 11th century AD composed by Jayadeva, a saintly poet, is a mystical allegory of the moods of love of Krishna and Radha. Krishna, the human incarnation of the god Vishnu, is the patron deity of the flavour of love. He is a pastoral god, called Govinda or Gopal, the cowherd, the divine flute player, who dances in dalliance with the milkmaids (*gopis*). Legends of his childhood pranks, and teasing of the *gopis* are the themes of many beautiful songs that are enacted by dancers highly skilled in the use of gestures and the exposition of sentiments depicting the moods of love.

The *Gita Govinda* has been sung and danced in the Jagannath temple in Orissa for several centuries. Narahara Tirtha, a saintly devotee of Krishna, brought dancing girls from Kalinga (Orissa) to teach the *Gita Govinda* and its language of gestures to *Devadasis* (temple dancers) and *Raja-nartakis* (court dancers) at Srikakulam in Andhra Pradesh.

The *Gita Govinda* is held in great reverence throughout India. In the eighth canto of the *Gita Govinda* the love situation of Radha is rendered in expressive mime as follows:

And then – in weary vigil the night passed.
Forlorn Radha waited
Assailed by Kamadeva's arrows until the morning came.
With repentant words, the Beloved One appeared before
  her.
But she, in angry jealousy rebuked him.

Kahetrayya, a poet musician of Andhra, and a fervent devotee of Krishna, composed beautiful Krishna love songs. He was a highly honoured court musician at Tanjore (Tamil Nadu) in the 17th century. An expert in mime, he personally taught his songs to the *Rajadasis* (court dancers) of the Tanjore Court.

169

Folk Dancers from Gujarat add colour the Republic Day parade in New Delhi.

Varnam is the most complete, versatile form of dance and mime set to music of a high order in the *Bharata Natyam* repertoire. Dramatic dance, interpreting each couplet of the song, is rendered alternately with pure dance, and, in the final theme *(charanam)*, the combination of gestures with dance provides a brilliant climax. *Varnam* begins with a pure dance prelude rendered to recited rhythm syllables in slow, medium, and quick timing, accompanied by the drum and recited by the dance director. Musical sequences of *swara* patterns (*sa-re-ga-ma*-etc.) accompany enchanting variations of dance interludes.

The most captivating element of the *Varnam* is the exposition of poetic meaning beyond the context of the song that express the moods of love in different gesture themes. This is called *sanchari bhava*. An example of this is as follows:

Hear me O Lord Siva, dweller in the fair city of Tanjore, I am completely depending on you, believe me.

This is rendered in dramatic dance first according to the text. Then, while the singer continues to sing the same line of poetry repeatedly, the dancer creates gesture and mime sequences of her own, such as:

As the bee depends on the lotus, believe me, by fire:
As the fish depends upon the ocean, believe me, by heaven and earth!

The *Bhagavata Purana* of ancient lore that extols the worship of Vishnu and Krishna contains many stories of the lives of these deities that aim to direct mankind to the spiritual path.

These stories were given dramatic form as dance-drama by Brahmin devotees in Andhra. The plays were staged at temple festivals by itinerant groups of *Bhagavatas* (holy devotees) in order to popularise religion. This art was known as *Bhagavata Mala.*

The village of Kuchipudi was entirely inhabited by these *Bhagavatas,* and they are still performing these dance dramas. This dramatic art seems to have existed at Kuchipudi from about the 13th century AD, and was later named *Kuchipudi Natya.*

Kuchipudi dance-drama includes speech, song with dramatic mime and dance, pure dance, and theatrical costume and make-up. All roles, male and female, are acted by male actors. The gesture language and exposition of sentiments are according to the *Natya Sastra* tradition. The *Bhagavatas* have a very precious repertoire of classical songs that are rendered with dramatic dance, and dramatic solo dances that can be danced by women. These dramatic dances have captivated the professional exponents of *Bharata Natyam* in recent years, and are sometimes included in the *Bharata Natyam* performances.

Orissa is known for its ancient temples, unsurpassed for architectual grandeur and sculptural beauty. The temples at Bhubaneswar, Konarak, and Jagannath are profusely adorned with sculptured figures that reveal a rich panorama of cultural life in which the arts were highly developed.

A *natamandap* or dancing hall was an essential feature of temple architecture, and music and dancing became a part of the daily ritual. For nearly a thousand years temple dancers known as *Maharis*, were attached to the temples of Siva, Vishnu and Krishna under the patronage of ruling monarchs.

The rounded curves and flexions of the *Orissi* dance are seen in hundreds of beautiful dance sculptures that adorn Orissan temples. 'In Hindu art the ideal sculptural postures are based upon bends or flexions which represent deviations of the body from the central plumb line of equipoise. These bends are called *Samabhanga* (equal bend), *Abhanga* (slight bend), *Atibhanga* (great bend), and *Tribhanga* (triple bend). The *Tribhanga* or triple bend flexion is a recurring plastic motif in *Orissi* dance, with the hip set at an angle. The *Natabara Bhangi* posture, which begins the invocation dance to Lord Siva, is a *Tribhanga* posture. The *Atibhanga* flexion in *Orissi* dance is a rhythmic movement of the torso alternately to right and left with the knees deeply bent and the arms simultaneously extended to each side. The *Bharata Natyam* dancer bends the waist; the *Orissi* dancer bends the hip.'

The *Orissi* danseuse has a marvellous command over facial expression and gesture in rendering gesture-song. The *Gita Govinda* of Jayadeva, immortal love-poem of Radha and Krishna, has been danced and mimed by the temple dancers for the last four hundred years. Orissa, the birthplace of Jayadeva, has cherished this art.

171

Kathak dance takes its name from *Kathak* or 'story'. There are reference in ancient Hindu literature to the *Kathakas,* storytellers, who recited epics and sacred legends with dramatic gestures, song and dance.

It is the classical dance form of North India which flourished in the medieval period as a religious art. It came into prominence under royal patronage in the courts of the Rajasthan states, in a secular form in the court of the Mughal emperor Akbar, and later at the court of the rulers of Lucknow.

The Lucknow school of *Kathak* developed the expressive form of the dance according to the taste of the royal patrons. The love lyrics of Radha and Krishna and the interpretive love songs called *Thumris* composed by Nawab Wajid Ali Shah in the early 19th century gave importance to mimed gestures and facial expressions in the *Kathak* style as well as pure dance.

The efflorescence of *Kathak* dance after a long period of decline has brought about a fusion of the Jaipur and Lucknow schools – the former specialising in pure dance and the latter in expressive dance.

The basic style of *Kathak* is *Nritta* – pure dance. Rhythm patterns of the feet, set to rhythm syllables in various permutations of the timing cycle, are the technical formula of *Kathak*. The lyrical movements of the arms articulate the curves and pivots of the body creating beautiful plastic patterns of dance called *tukras* that end in a still pose exactly on the *sam,* the central emphatic beat of the timing cycle.

The *tabla,* a pair of drums, accompanies the dance supported by a musical refrain on *saraugi* or harmonium continually repeated to guide the dancer. The patterns of rhythm syllables are usually recited while the dancer marks time with the feet and then interprets the sequence in dance.

The stories of Krishna, the pastoral God, his dalliance with Radha and mischievous games with the *gopis,* is the usual theme of expressive mime with dance. The dancer alternately depicts the lover and beloved, passing from one role to the other by means of the *palta,* a spiral turn from side to side. A *Kathak* dance recital reaches its climax with *tatkar,* a display of complex footwork in multiples of the basic *tala* (timing).

Kathakali theatre, the dance-drama of Kerala, is an art of dramatic expression through the medium of gestures, facial expressions, and mimetic dance. The highly stylised facial expressions are accentuated by fantastic mask designs that are painted and moulded on the faces of the actor-dancers to intensify the movements of eyes, brows and facial muscles, and the theatrical spendour of jewelled head-dresses and colourful costumes create the symbolic personages of the *Kathakali* theatre.

The gesture language of *Kathakali* is a grammatically complete language of hand symbols, that not only serve the purpose of communication, but imaginatively interpret poetic meaning, sentiments, and dramatic action. Story-poems from the epics *Ramayana* and *Mahabharata* are sung by the vocalists accompanied by drums and cymbals. Scenes of challenge and combat are played on the high-pitched *chenda* with drumsticks. In the *Kathakali,* an open-air dance-drama, all the roles are danced by men; it is the richest, and most striking dance-drama of India.

India's traditional dance modes and gesture language, which are now dispersed among several regional forms of classical dance, provide a rich vocabulary of choreography and mime, linking the past with the present and bringing together divergent manifestations of the one classical tradition.

The Cinema in India

Amita Malik

According to United Nations statistics, India has one of the four largest film industries in the world. In annual production, it has never been lower than fourth, and it has been as high as second. It has long out-distanced Hollywood and has been left behind only by Japan. The cinema is the largest and cheapest medium of mass entertainment in India, and its most successful popular art. Statistics are seldom issued officially on these subjects, since much of India's film finance statistics lie buried, but an assessment made some time ago rated it as the second largest industry in terms of capital investment and the fourth largest in terms of labour employed. Indian film stars have been known to influence voters by public appeals on the streets of Bombay to an extent that popular politicians could not. In times of distress or natural calamity they have acted as fund-raisers whom no-one, from pavement-dwellers to millionaires, can refuse, and they exert an amazingly calming influence on highly exited mobs or lonely soldiers. The film in India has not only brought an entire new system of social values into Indian life, it is also radio's and now television's, most formidable rival and unlikely to be put on the defensive by other mass media, since its first aim is entertainment and it has a unique personality of its own.

India is one of the few countries where the government is actively participating (some call it interference) in film-making, leading to strange twists in censorship and other moral slants. But, equally, the government has made several constructive contributions to the more serious side of film-making and has been very much part of the film scene.

When the Lumière brothers of France held their first bioscope show in Bombay in 1896, they could hardly have anticipated that they were assisting at the birth of India's most

174

powerful twentieth century folk art. It was a simple enough show of two films: *Arrival and Departure of a Train* and *Bathers in the Sea*. But so great was the demand for seats at the Elphinstone Theatre that soon a large, separate box had to be provided for *purdah* (veiled) ladies to come without their husbands.

And, indeed, the arrival of the cinema in India could not have been better timed. It was the turn of the century and city audiences were hungry for mass entertainment. British repertory companies playing *Hamlet* were all very well for the Westernised élite and fiercely nationalistic indigenous plays such as *Neel Darpan* (Blue Mirror), dealing with the exploitation of workers on indigo plantations, also filled the professional theatres. But for uneducated city audiences, a substitute had to be found for folk theatre and the morality plays which still formed such an intrinsic part of village life and which had been undisturbed by British rule.

For these mass audiences the cinema, with its direct visual impact, its easy accessibility and its simple themes, seemed the natural answer. And now, in the 1970's, when about 70 per cent of India's over-500 millions are still illiterate, the cinema remains the largest and cheapest form of mass entertainment.

The tremendous artistic and social possibilities of the cinema were, however, grasped in its earliest days by a young engineering student studying in Europe. Dadasaheb Phalke, later to be called the Father of the Indian Cinema, was deeply moved by a film he saw in London on the life of Christ. Its parallel with the life of Lord Krishna was as striking as its cinematic force. Phalke, on his return home, plunged bravely into film-making. He once sold his wife's jewellery (like his distinguished successor Satyajit Ray) to finance *Raja Harischandra,* India's first feature film, made in 1913. Its plot, derived from ancient mythology and known to every Indian, was to prove symbolic: Raja Harischandra, the good king, renounces his throne and goes into banishment in the wilderness after a promise made to a holy man. In Indian morality plays, too, Good always triumphs over Evil, but only after much suffering and spiritual struggle. *Raja Harischandra* proved an instant success, and Phalke, an improvisor and technician of undoubted genius, lived long enough to see that the safest box-office film in India is the mythological film. It has survived everything, from puritanical censorship laws to James Bond.

At a time when there were social taboos on women in the theatre and boys played women's roles, Phalke bravely enlisted his own wife and daughter to take part in his films. His artistic integrity was as courageous as his professional boldness. He was also far-sighted enough to leave a film record of himself at work which is the most treasured film in India's National Archives.

Meanwhile, in 1911, an enterprising American, Charles Urban, had shot the first documentary film in India, *King George V and Queen Mary at the Durbar.* It inspired the more nationalistic Indians, at a time when British censorship rendered impossible the making of political films, to keep a detailed record of the freedom movement. When the Gandhi Centenary year was celebrated in 1969, it was found that cameramen from all over the world had contributed to a five-and-a-half hour documentary film on Gandhi, covering every conceivable period and setting: the passive resistance campaign in South Africa in 1912; Gandhi with goose-stepping little boys in Mussolini's Italy; rare shots with Chaplin and Bernard Shaw in London; and his lonely campaign for communal harmony in Noakhali in former East Pakistan when Delhi was ushering in Independence with pomp in 1947.

Unlike the Japanese cinema, which burst upon a startled world with *Rashomon* at Venice after World War II, the Indian cinema was deliberately internationally-minded from the start.

A barrister from a distinguished Bengali family was also drawn to the cinema while in London. Himansu Rai made *Light of Asia,* on the life of Buddha, in 1923, with a German cameraman, Frank Osten. It was a great success in Europe and England in the early twenties. Later, with his actress wife Devika Rani, he made his best-known international success, *Karma,* with English dialogue. It ran at a London West End cinema for ten months and won the *Daily Express* prize for the best film of the year.

Rai met Devika Rani, a great-niece of the poet Rabindranath Tagore, in London when she was seventeen and doing a course in stage makeup. They worked for some years with UFA studios in Germany, where they were contemporaries of Dietrich and Fritz Lang. Returning to India, they set up the Bombay Talkies, a studio organised along modern lines. They insisted on fresh young talent, preferably from the universities, and discovered a whole new generation of film-makers. The Rais chose the best in Indian culture and tradition to tackle burning social issues. In *Achyut Kanya* (*The Untouchable Girl*) they showed great courage in attacking Untouchability. Audiences all over India were deeply moved by the sequence where a village girl is ultimately stoned to death for daring to love a young man of high family.

The thirties, the golden age of the Indian cinema, was also the time when powerful social themes were put across with refinement as well as popular appeal. In Bengal, New Theatres, with socially and artistically dedicated film directors like Pramathesh Barua, drew on the novels of Sarat Chandra Chatterjee and other writers to make films of powerful social impact. They made films of the saints who sang religious songs and the actors and actresses of those days included singers who needed no dubbing.

In Maharashtra, in large, well-organised studios such as Prabhat and Ranjit Movietone, artists worked under conditions of artistic freedom as well as professional security. Maharastrians excel at social satire and used the cinema to expose social cant and hypocrisy. Stage and screen talent were frequently interchanged and there was no lack of playwrights for both.

Above all, there was no gap between highly educated and illiterate audiences because it was an age of artistic idealism when themes were constructive and contemporary and also ageless. It was a refinement of folk theatre with a little sophistication added, but no compromise in basic Indian values or cultural heritage. Their Westernisation was an asset for mass communication skills and gave them an intellectual detachment when facing the uglier, more emotion-charged social problems. Theirs was a cinema of involvement.

The problems of India's cinema today are a direct result of World War II. Wartime conditions had already crippled the great studios. The final blow came from the *nouveau riche* investor of the post-war period. The end of the war saw the advent of the war profiteer who had moved into the cinema and changed it beyond recognition, sensing a heaven-sent field for investment, with its guaranteed audiences and sure profits. So great was the flood of finance that it swept all, including the well-established studio system, before it. It also ushered in the age of the freelance film-maker with more money than talent. New terms like 'box-office formula' and 'the star system' followed, as did that uniquely Indian film phenomonen, the 'playback singer'. The film director and story writer were pushed completely into the background. The singing saint was replaced by mammoth concert orchestras, Hollywood style, accompanying a new type of film pop music, catchy lyrics dubbed on to the new-style hero and heroine as they chased one another round the bushes. With the new 'boy meets girl' box-office formula there emerged an oblique form of pornography to by-pass the censors: belly dances by night-club performers and folksy group dances with ill-concealed sexual overtones. The number of songs and dances in a film became far more important than the story. The fees of stars now ran to astronomical figures for each film and they appeared in five or six simultaneously, so great was the demand. The steady monthly salaries of the old studios now seemed laughable. For the new movie moguls were not only laying down artistic standards – they were prepared to pay for them.

The 1950s saw the establishment of what the rest of the world rather vaguely calls the 'typical Indian film'. This is a marathon, which Indian audiences treat very much like a picnic, a day's outing with the family, complete with packed lunches, babies wailing in the auditorium, and an audience which includes wives of industrialists in their newest imported nylon saris to college students, illiterate domestic servants, peasants who cannot read the credit titles, and taxi drivers who are happy to miss half a day's earnings to witness a personal appearance by a popular star.

The typical film plot has something for everybody, since it is, in effect, a tragi-comical-musical-historical-sociological dance-drama which audiences in developing countries in Asia and Africa devour wholesale, and quite often without sub-titles or dubbing, so strong is it visually, and so familiar the dialogue in any Eastern language. The overseas fan mail of All-India Radio runs into thousands of requests for film songs. One listener in Tokyo confessed he could not sleep at night unless he heard an Indian film song before going to bed. Indian television's peak viewing time is over weekends, when it shows commercial feature films. Radio and TV are not only the cinema's poor relations, but also largely dependent on it for their most popular programmes.

But in the typical film, which the film societies and film festival audiences both in India and abroad choose to ignore, two fundamental aspects of the Indian scene remain intact. And, ironically enough, invest the Indian cinema with its unique split personality, for two completely different, highly complex and conflicting types of film co-exist: the Bombay-Madras glossy and the thoughtful, idealistic regional film. It makes the Indian cinema at once the most confused as well as the most sociologically interesting in the world, and this affects national as well as foreign audiences brought up on Satyajit Ray.

The mass-produced, glossy Hindi film, with excellent and highly modern technical values, passes through periodic financial crises. It also causes frequent government income-tax investigations – 'black' money passes under the counter in star fees. Madras, with its well-organised and professionally orderly studios, has now surmounted the language barrier and is flying down stars and production teams in plane-loads from Bombay to make big-budget, tear-jerking Hindi films which have recently proved more successful at the box-office than Bombay's. And this, while language riots and boycott of cinemas take place in both cities from time to time against infiltration of cinemas by language from the other hated region.

For in its crude, commercially minded, ham-fisted way, the Bombay film and its kin still remain India's most popular folk art and the cheapest. And it has a good deal of horse-sense as well as Indian-ness under that terrible veneer. The classical Sanskrit play also had elements of song and dance, like the Bombay film. It also had its comic interludes. So the song, dance and comic elements which so amuse and irritate the sophisticated can no more be dismissed as un-Indian than

179

the hammy villain with the false beard seen in every morality play in every Indian village at festival time.

The basic film plot, however escapist and garbled, still pursues the triumph of Good over Evil. But now Evil is represented by the belly-dancing gangster's moll in the nightclub, highly Westernised and wearing a blond wig. Good is represented by the traditional Indian woman, devoted and long-suffering, whose gullible fiancé or husband is lured to crime by the belly-dancer. Even if the villainness now chases her man round and round the Eiffel Tower, Niagara Falls, an Olympic stadium or the swimming pools of Beirut, the audience knows that the hero will see the light and return in time to light his wife's funeral pyre. The hero's sports car, bottle of whisky and chintzy modern apartment – straight out of the glossy magazines – might be miles away from the factory worker or garbage collector, but it is an enticing form of escape after a hard day's work.

In this seductive context, the regional minority cinemas of India face a struggle, because the Bombay film has infiltrated every town and village and dominates the city cinemas. But their minority status is only quantitative. The regional languages of India have rich, proud and highly intellectual traditions and writers actively in touch with contemporary problems. Some towns, such as Calcutta, have had established professional theatres running continuously since the 1850s. They therefore have a steady clientèle from the educated middle class which has always patronised the theatre. They are quite often the leaders of political and social thought. For them, an intelligent small-budget film which provokes thought and argument is not only good cinema, it can also be good box-office. The regional film-maker can face such audiences without publicity gimmicks and expensive cocktail parties, because he knows where he stands with them.

When Satyajit Ray made *Pather Panchali* (*Ballad of the Road*) in 1955, he was exposing in universal terms and with unique artistry the living problems of his own people. Based on a Bengali novel, it confronted the problem of poverty and the educated high-caste village family. Later *Mahanagar* (*The Great City*) tackled the problem of urban poverty in the middle classes and the domestic upheaval in a traditional home when a young married woman takes up a career because the men cannot support the family.

In Ray's *Kapurush* (*The Coward*) there is the bored, affluent planter's wife and the domestic crisis when the young man who did not marry her encounters her equally bored husband. *Charulata*, perhaps Ray's most exquisite film, concerns an editor who, because he is obssessed with his work, neglects his beautiful and intelligent wife. A period piece by Tagore, it is set at the turn of the century, but is as contemporary as it is universal. *Jalsaghar* (*The Music Room*) tells of the decay of the aristocracy as typified by a lonely old man who rides proudly to his death after a last, defiant, show of grandeur. In *Pratidhwandi* (*The Rival*), the study of an unemployed boy in Calcutta, Ray again deploys his sensitive understanding of modern social problems with artistic finesse.

These are problems known and understood by every Bengali, and which yet find an echo throughout the world. Ray's audiences accept the understatement, the subtlety, with as much naturalness as the Bombay factory worker accepts the Westernised girl who sings 'Happy Birthday to You' round a Christmas tree.

Mrinal Sen, another Bengali film director, has made *Interview*, again with an unemployed youth in Calcutta as its subject, with the same staccato technique as his *Bhuvan Shome*, a study of an arrogant bureaucrat humanised by a village girl. Also in Calcutta, Ritwik Ghatak, a lonely intellectual, has made several films with unusual themes (such as the love of a tough taxi driver for his vintage car), and Tapan Sinha, more in the classical Bengali tradition, has made perceptive films on modern youth and general social problems.

If Ray, Sen and Ghatak succeed outside Bengal, it is partly because of the film society movement, which has made Indian audiences much more nationally and internationally minded. As the founder of the Calcutta Film Society before he made his first film, Ray's contribution to the awakening of Indian audiences is as profound as the films he has made. Membership of film societies in Calcutta, Madras and Bombay now run into thousands. More significantly, there are film clubs in the oil towns of Assam, the steel towns of Bengal, and in the small pilgrim towns in the Kumaon hills. The country is now

181

thinking in terms of art theatres which will screen both Indian art films and imported art films not commonly shown in commercial cinemas.

In India, the documentary film has a tremendous role to play in mass education. As one of the largest producers of documentary films in the world, India has a government-sponsored Films Division which releases one documentary film and one newsreel every week in commercial cinemas. Showing these is compulsory and every cinema contributes a small percentage of its takings for the two-reeler film. While this might seem dictatorial, at least release of Indian short films is in Indian hands, and it has stimulated a strong documentary movement as well as a feeling for documentaries among filmgoers in India. The independent documentary film producer in India, who had to fight tremendous odds before, now finds government sponsorship. The new wave in Bombay started with documentary film-makers such as S. Sukhdev (*Miles to Go, India 1967*), V.S. Chari (*Face to Face*), M.F. Hussain, one of India's leading painters, whose *Through The Eyes Of A Painter* won the Golden Bear in Berlin. In the Films Division itself, there are documentary makers who quite often surmount bureaucratic curbs to make films dealing with contemporary reality. They are now being joined by a new generation of feature film makers in Bombay itself, and a whole new generation of highly professional internationally-minded film-makers is emerging from the government-sponsored Film Institute in Poona who are as much at home with Raj Kapoor as with Bunuel, Antonioni and Peter Brook. They are already defying the movie moguls and have scored at the box-office with small-budget black-and-white films. The Film Finance Corporation of India, also government-sponsored, quite often finances these films, even taking risks with young, unknown film-makers who could not possibly raise money from the money-lenders who finance films of the box-office variety, often at high rates of interest.

Two new-comers to the cinema are working in western India, in Gujarat – India's industrial belt; N.G. Saraiya, who trained as a cartoon film-maker with Norman MacLaren and the National Film Board of Canada and made a film called *Saraswati Chandra* which established him in feature films, and

Kantilal Rathod who made *Kanku*, his first feature film in Gujarati, the courageous story of a pregnant village widow. It won its actress, Pallavi Mehta, a Best Actress award at the Chicago Festival in 1970.

A significant trend comes from South India, where films play a vital role in politics. In Madras, in the struggle between the Congress Party and the DMK Party, some leading politicians, including government ministers, are top film writers. Stars belong to different political camps and their roles on the screen quite often take on powerful significance in public life. Subliminal nuances in colour films have been used to great effect. As red and the rising sun are party symbols of the DMK, whenever the sun rises or the decor of a film is red, there are thunderous cheers or boos from the audience.

In the South Indian state of Kerala, the first to have a legally elected Communist government, audiences are the most highly literate in India with sharp political, social and literary awareness. Here there is a solitary film-maker, Ramu Kariat, whose first film, *Chemmeen* (*Prawns*), based on a prize-winning novel by a Malayalam writer, won the State Award in India. It also found a response in places as far removed as Chicago and Eastern Europe.

With the regions in ferment, the gap between the Hindi and the other language films is becoming perceptibly narrower, and Bombay is already a little on the defensive, especially as it faces periodic production, distribution and exhibition crises of a crippling financial nature.

In this context the burning issue, and, indeed, the perpetual censorship issue, remains: to kiss or not to kiss on the Indian screen? It is hotly debated in homes, coffee houses, behind the sacred portals of the Indian Government and on university campuses. While the traditionalists say 'no' (after all, they assert, Indians never kiss in public, so why on the screen?), the progressives are in favour, especially where the aesthetic and social situations warrant it. The highly controversial report of the Khosla Committee, set up in 1968 under Mr Justice Khosla to conduct an enquiry into film censorship, also agrees with enlightened public opinion on this point. It aptly sums up the dichotomy not only in Indian life, but in its cinematic mores.

183

With its basic traditional values, the Bombay film is likely to survive only in more chastened form as finance becomes less fluid and values more liberal. The Bombay film still adds much colour and gaiety to the drabness of life for the average Indian. As audiences become more discriminating with better education, it is likely to become more sophisticated, but, possibly, much less fun.

The remarkable thing about the Indian cinema is that its two aspects can co-exist so peacefully, and that Satyajit Ray can surmount as well as survive the Bombay dream sequences.

Left: *A canopy in the Meenakshi Temple complex in Madurai. The chain link hanging from the corners are carved from solid stone.* Right: *The Pearl Mosqu in the Red Fort in Delhi, seen from the Hall of Special Audience.*

Food in India

Robin Howe

Food holds an important place in Indian life. Hindus believe that food was created by the Supreme Being for the benefit of man; so it follows that cooking, serving and eating food must be a sacred ceremony attended with religious ritual. Because religion permeates all aspects of daily life it is the intimate relationship between man and nature in India which determines the Hindu's attitude towards the food he eats. Regarded with reverence, for him food is the secret of happiness and of human relationships.

Almost every discussion in India is prefaced with: 'India has had five thousand years of civilisation'. An examination of food is no exception. Food, its value, its joys and its preparation, has considerable mention in the Hindu sacred books, the *Vedas*, the *Upanishads* and the *Sastras*, the last a series of books compiled sometime in the 4th century BC. These books, concerned mainly with the spiritual side of Indian life, have also dealt with considerable care with the question of health and feeding. Stringent rules concerning food, its preparation and eating, especially in regard to hygiene, were laid down. So many generations of Hindus have followed these tenets that they have become part of the Hindu way of life and inevitably part of the Hindu religion.

The vast majority of Indians are vegetarian, but food habits change from caste to caste and province to province. Although there are four main castes, there are some three thousand sub-castes and almost every Hindu has his own idea of what constitutes vegetarianism. In addition to the accepted refusal of meat and poultry, some add eggs to the list, for these are life in embryo. Many vegetarians do not eat cheese because of its rennet basis.

Not only religious and caste restrictions have affected the

185

Kathakali dancer. The present form of the dance is believed to have evolved between the 15th and 17th centuries A.D.

eating habits of the Indians; the weather, geography and the impact of foreigners all have played their part. Where the sun and the climate together produces an abundance there is a relaxation of the rules to enable the Indians to avail themselves of this bounty. For example, fish is eaten by the Bengalis and the Brahmins along the Malabar coast, suggesting common-sense rather than religious tutelage. In the south where the soil is not so fertile and the rainfall spasmodic there is less variety of produce than in the Punjab. So the southern Indians were, generally speaking, orthodox and have remained that way. It is hot all the year round so that it is no great privation to abstain from meat-eating – even Westerners and Indian Christians find it too hot to eat meat all the time.

In the north, the home of the sturdy Punjabi, the weather varies from a dry scorching heat to intense cold but favoured with winter rains. Here the food is much heavier and richer. Here, too, the Muslims have penetrated, bringing with them their meat-eating habits, their rice dishes of *pilaus* and *biryani*, stuffed sheep and goats, so that today all this is part of the Punjabi diet. Because also the quality of food in this northern region is good, much less use is made of chillies and spices.

But whatever dietary laws exist, most Indians would not consider life worth living if they had to eat totally unspiced food. To all of them the food of the West is insipid and without flavour.

In principle, food is divided into two main groups: *pukka* (good) which means certain foods cooked in pure *ghee* (clarified butter) or in curd. The second category is *kutcha* or 'poor' food, which means the meat of most animals and the birds of the jungle. It also embraces such items as snails, frogs and beef.

The *pukka* category is sub-divided into *Satawick* and *Rajasik* foods. The former is intended for Brahmins and allied castes and classified as conducive to health and spirituality. It includes parched grains, fruit and most vegetables but not onions, garlic or mushrooms and several of the root vegetables.

Rajasik food is much more liberal and allows the eating of almost everything except beef and garlic. This exclusion of garlic is odd but the earnest compilers of the *Sastras* considered that this vegetable, beloved of the Latins and Asians alike, roused men's passions and baser instincts. Over the centuries this edict has been forgotten for there is a lot of garlic in Indian cooking today. But the *Rajasik* diet was wise and well conceived. It was designed for the temporal leaders of the people, the warriors and princes whose requirements were aimed at manhood, strength and power. These were the men of the Kshatriya caste. Their diet included meat, wild boar but not pig since the latter is a dirty eater and its flesh putrifies swiftly in the heat. Fish was also forbidden, a matter of sense since it is a bad traveller and could not be kept fresh in those days of non-refrigeration.

Not only must the high caste Hindu consider carefully what he eats but also who cooks for him. A man who cooks or eats low caste food is also low caste himself. Fortunately not all high caste families are men of wealth and learning so there are Brahmin cooks who may cook for their own caste, allied and Brahmin sub-castes. But always with reservations, mostly concerning the use of their own pots and pans. There are many non-Brahmins who keep two cooks, one to take care of their Brahmin guests and another for their own cooking and that of non-Brahmin guests. Naturally this involves two kitchens. I have known a college in the south which had eight kitchens to cope with the different castes of their pupils and their food restrictions.

This division of eating stems from the days when Brahmins were really the spiritual leaders of the people. To set an example and to be purer than other men they had to eat the sort of food adjudged good for the mind. People of other and lesser castes did not have this responsibility but then, as now, there was an aping of one's betters, a sort of spiritual keeping up with the Joneses. Hence the present confusion.

Very few Indians drink alcohol, the bulk of the population practising voluntary prohibition, although drinking was not forbidden in the ancient scriptures. According to the Laws of Manu, one of the ancient saints and writers: 'No sin is attached to the eating of flesh or to the drinking of wine'. But, say the more austere, and here is a big but, neither is conducive to the Hindu way of thinking and the spiritual life. However, in some industrial centres and in the villages workers and peasants like to drink toddy, a very alcoholic drink tapped from the toddy palm and actually recommended by dieticians as supplying useful vitamins which their diet of rice and lentils lacks. The Westernised Indian who drinks prefers what is officially called 'foreign liquor'.

Tea is almost the national industry of India, also the national drink except in the south where coffee takes precedence. 'Our Brahmin brew of coffee will not yield place even to the most carefully prepared *café-au-lait* of France', wrote one South Indian writer. Soft drinks are many; excellent buttermilk made of the curd of rich buffalo milk; *lassi*, which is soured milk mixed with water, sugar and crushed ice and is deliciously refreshing; *rasam*, a highly spiced pepper-water, is favoured in the south which, while it takes the roof off one's mouth is by no means displeasing. There is *nimbu pani* or fresh limejuice sweetened and diluted with water and one of the finest-ever hot weather drinks; a vast number of sherbets, syrups and cold fruit drinks in which all available fruits are utilised. There are milk drinks, often flavoured with rosewater and almonds and sometimes rather gaudy. Where the coconut flourishes there is ice-cold liquid from the young green coconut, truly one of the ambrosial drinks of the tropics.

Many are the extravagant claims made in favour of Indian cooking, its richness, its subtlety of flavour, its high place in the Indian scheme of things and its diversity.

Cooking is an ancient art and many of the Indian dishes require skill and considerable patience to prepare. For many official dishes cooks must spend all day cutting up vegetables, simmering milk or pounding rice and spices on a square stone slab or sitting patiently over a fire for hours on end. On the other hand there are the simple dishes for daily eating. The boast that a good Muslim cook can produce a hundred different mutton dishes is probably true. There are many varieties of kebab, some fried, some grilled, some stuffed with nuts and others with cream. Equally popular are the *quoormas*, a sort of heavily spiced dry ragoût. Kid, goat and sheep as meats are interchangeable. In Hyderabad there is superb Muslim cooking, all of which adds to the richness of Indian food.

A northern dish which almost all foreigners favour is *tanduri* chicken which takes its name from the oven in which it is cooked. The plucked chicken is rubbed in curd and spices, left for several hours then roasted inside the *tandur*, a primitive clay oven with a wood fire burning fiercely beneath. The chicken hangs over the fire from an iron rod and is served hot. It is eaten with the fingers accompanied by slabs of *nan roti*, a type of bread also made in the *tandur*. It is extremely good when eaten absolutely hot and fresh. A large plate of chopped onions, laced with chillies is served as a side dish.

Now for curry – the choice is wide. The origin of the name is obscure. One version is that it comes from the Tamil word 'kari' meaning sauce. It also means spiced food and the first 'modern' mention of curry is in 477 AD. Various travellers over the centuries have commented on the Indian way of eating rice and curry, including 'Ambassador' Atheneus, citing Megathenes: 'Among the Indians at a banquet a table is set before each individual... on the table is placed a golden dish on which they first throw boiled rice and then add many sort of meats dressed after the Indian fashion.'

Eating from a *thali*, a round tray, has not changed over the centuries, although the gold and silver has been replaced by cheaper metals. Sometimes the *thali* is placed on a low table and one sits cross-legged on the ground in front of it, shoes removed and feet tucked away. Or one sits on a low stool, the *thali* on the ground, or there is a table and stool; all variations on a theme. In the south both the *thali* and the flat shining green banana leaf takes the place of the western plate. The latter is thrown away afterwards thus saving washing up.

Around the inside rim of the *thali* are arranged small bowls of matching metal, each filled with a different sort of spiced, vegetarian food; potatoes, cauliflower, aubergine, beans and cabbage. One bowl holds a thin curry sauce, another thick curd, yet another a sweet concoction. In the centre of the *thali* is placed a heap of rice, a number of Indian pickles and chutneys, often a banana and some dried chillies. As one empties the small bowls they are quickly replenished by the women of the household who still, in many Indian homes, eat separately from the men.

Because Indians prefer to eat with their fingers they always

189

wash their hands immediately before and after eating. Eating with a knife and fork is unsuited to Indian food and there are piping-hot *puris* (fritters) or *chapatis* to scoop up the more liquid food. It is generally felt that cutlery cannot be properly cleansed, also I was told one has to test the heat of the food with the fingers. 'Food must be inspected, first with the eye, then with the fingers, after this the mouth, the tongue and finally the taste.' There is a finesse in eating with one's fingers, and with the right hand only, never letting the food come above the joints of the first two fingers or thumb.

At the end of most Indian meals *pan* is served. At a formal meal the servant brings round a silver salver piled high with small triangular bundles of spices as an aid to good digestion. *Pan* consists of a dark green leaf smeared with a lime or crushed rose petal paste wrapped around crushed betel nut and a variety of spices such as cloves, aniseed and cardamom, as well as grated coconut. For special occasions a coating of silver leaf is added.

When neither the *thali* nor the banana leaf is used then eating is basically Western style. There will be a platter of rice, always perfectly cooked and a number of curries; *dahl*, which is medium-thick lentil purée or soup, and chutneys which are freshly made and include a dish of curd with chopped cucumbers or tomatoes, a paste of chopped mint and grated coconut, tiny green pickled mangoes and possibly some pickles.

In the Punjab, cooking is done mostly in *ghee* or butter, except for fish which almost all Indians cook in oil. And the type of oil differs tremendously. In the south it is always coconut or sesame oil, in Bengal mustard oil, and this applies to much of the north also. The Punjabi meal will usually consist of meat dishes and curries which include a cheese called *panir* which resembles the Greek *feta* cheese.

There are splendid *pilaus* 'fit for a Maharajah's table' garnished with fried onions, sliced hard-boiled eggs and toasted nuts and finally topped with shimmering gold or silver leaf – for festive occasions also scattered with rose petals. One is reminded of Emily Eden's observation in *Up the Country*: 'The Rajah sent us some dinner – three kinds of roasts and covered with gold and silver leaf, a deer and about fifty other dishes of sorts.' Those were the days.

Another Punjabi speciality is *khoya*, a cream so thick that it can be grated. Justly renowned are Punjabi stuffed *parathas*, and their roasts and skewered dishes; they also claim the finest lentils of which there are over sixty varieties. These are called the poor man's meat for they are rich in proteins.

For many Indians the cooking of the Punjab is the *cordon bleu* of Indian cooking, a claim which causes the man from Bengal to froth at the mouth. He considers that the Bengalis are the élite of the country, their cooking the most exquisite, offering the best balanced diet in the country. He points to the plethora of fine fish dishes – one of the most successful fish dishes in Bengal is the *hilsa*, fish delicately spiced and wrapped in pumpkin leaves for cooking. To achieve variety the Bengali cook goes to the flowers and unusual fruits, and one of their specialities is a curry of bamboo shoots.

And, of course, all Bengali cooking recipes are secret, handed down through the centuries. A cook who has been trained to match the exacting demands of a Bengali's taste buds is for the Bengali a treasure not to be lightly cast aside. Despite all this it will be difficult to find a Bengali restaurant; to eat Bengali food you must cultivate Bengalis and get them to invite you home.

South India is another matter altogether. It produces the most highly spiced food of the country. In almost every dish there is something from the coconut palm. In Kerala, the southermost Indian state (the name means 'Land of the Coconuts') a coconut palm grows in every back yard. The South Indian cooks in coconut oil, uses coconut grated and in chunks, drinks vast quantities of *neera* or coconut water and considers no religious ceremony complete without a piece of fresh coconut as an offering.

Cooking in the south is almost entirely vegetarian, with a meal starting and ending with rice. It is, in many ways, somewhat simpler cooking than that of the north or west but on the whole the kind of food which has earned the remark: 'the climate is hot, the dishes are hotter and the condiments are hottest'. Many is the Western palate which has reeled at the first bite of a South India curry yet has come back for more. This is also the province of tropical fruits, some so lovely they

191

deserve only to be served on the finest porcelain.

On to Bombay on the western side, where the food more often seems to be a happy combination of east and west. Both rice and wheat are included in their diet and although most Maharastrians and Gujaratis are vegetarians there are many who include meat, eggs and fish in their diet. Fish are plentiful along the coastline and modern fishing methods and refrigeration are making it possible for fresh fish to reach most parts of India. Bombay prawn curry is something to remember, so is their pomfret.

In Bombay live the Parsis, or most of them, adding their quota of rich dishes to the culinary scene. Further down the coast is a Portuguese influence with sweet-sour dishes like *vindaloo*, sophisticated curries like duck *bafad* and mild ones such as the egg or chicken *moolie*.

Despite all these differences most foreigners still think of Indian food as curry and rice, searing hot and acrid. A curry is only as hot or spicy as the cook chooses. It is not a greasy mess, nor should it be garnished with apples, sultanas or raisins, any more than a dish of *kicheri* (kedgeree) is a sticky rice dish mixed with fish and served for breakfast. It is true there are curries so hot that the unfortunate tyro dashes from the table groping for water. Indeed it is better to eat a banana or polish off a bowl of curd. Good curries, and there are many, are like good jokes; some are mild and pleasant, others spicier but still palatable.

Most of the spices used in Indian cooking were chosen originally for their medicinal qualities rather than any thought for flavour. There are still few ailments that the Indian mother – and often father – does not think can be cured with spices. Many of them such as cloves and cardamoms are very antiseptic, others, like ginger, are carminative and good for the digestion. Turmeric, one is told, is splendid against skin diseases, bruises and leech bites; neem leaves, rather more a herb, are used to guard against smallpox, while singers chew tamarind leaves to sweeten their voices.

Spices are bought as and when required. I have never heard of an Indian cook using packaged curry powder. He goes to the bazaar and buys just enough to make his curry, then he brings his choice home and grinds. The connoisseur of curries knows that in each curry go different spices, and few cooks use the same quantity or variety as another. Therefore, no two curries taste the same.

So, after all the historical and ritual side of Indian cooking, the differences between north and south, east and west, the vegetarian and non-vegetarian, I feel that it is the very complexities of the rules and regional differences which make Indian food so fascinating. Those who approach Indian food without prejudice will find it delightful and unusual, and worthy of the adage: 'There are three great kitchens of the world, the French, the Chinese and the Indian.'

Wild Life in India

E.P. Gee

There is evidence that there was some sort of elementary appreciation of wild life in India in the olden days. For instance the treatise on Statecraft called the *Artha Shastra*, attributed to Kautilya about 300 BC, contains a reference to certain forests which were specially protected 'with game beasts open to all'. Such forests were called *Abhavaranva*, and in some measure can claim to be the forerunners of the national parks of today. And in 242 BC the Emperor Asoka's famous fifth pillar edict gave protection to fish, animals and forests.

It is tantalising nowadays to think of the proliferation of wild animals and birds that must have existed in the olden days, and of such vast areas of unspoilt forest and marshland. Even as late as the reign of the Mughal Emperor Akbar (died 1605) there must have been very large herds of blackbuck in northern India for he kept as many as a thousand trained cheetahs for the sole purpose of hunting these antelopes, which are now quite rare.

During the past couple of centuries or so, with the British Raj established in India, big game hunting seems to have become a major pastime of both princely and British India. In practically every part of the subcontinent there seems to have been an almost unlimited amount of big and small game to be shot, and sport seems to have been as popular with the princes as it was with the British military and civilian officers and the European community in general. Truly India may be considered to have been a 'sportsman's paradise'.

With the improvement of firearms and ammunition, and with more and more people going in for big game hunting, it soon became apparent that some sort of regulation of sport would become necessary. Shooting Rules, Closed Seasons and the like, together with a code of conduct and ethics of

194

sportsmanship came into being – especially in British India which was more disciplined than the rest of the country. But in spite of this 'control', the slaughter went on almost unabated, and all the old *shikar* books testify to the amount of carnage carried on by officers and others of the British Raj in the name of sport.

Take the case of the Indian lions. These existed over most parts of western and northern India, north of the Narbada river, and a favourite pastime of officers was killing them either by shooting or by riding them down on horseback and spearing them. In the middle of the last century a certain 'sportsman' shot over 300 of them, fifty of which were in the neighbourhood of Delhi. By the 1880s there were no lions left in India, except in the Gir Forest in the old princely state of Junagadh. With such a sustained rate of killing, the stage was reached when some sort of preservation was necessary. In the provinces of British India this was achieved by the Forest Department's creation of reserved forests in which shooting was controlled, and a certain number of game sanctuaries in which no shooting was allowed but in which game animals could multiply and spread. Game preserves were also set up in most of the princely states so that game animals and birds could breed and sport continue unabated. Shooting inside these preserves was limited to members of the ruling family and very important guests; otherwise they were strictly protected from shooting by ordinary or unauthorised persons.

This protection of 'game', as it was then called, made a very valuable contribution to the preservation of wild life as we know it today, although sport rather than conservation was the main objective. The outstanding example, of course, is the Gir Sanctuary where the former rulers of Junagadh safeguarded the last stronghold of the Indian lion – but for this the lion would certainly have become extinct.

With the coming of independence in 1947 all the princely states acceded to the newly created Indian Union. As far as wild life was concerned the status quo was fairly well maintained in what used to be British India, although more freedom was allowed and more gun licences issued. But in the erstwhile princely states the authority and prerogatives which had been exercised by the rulers ceased, and overnight the people took over and helped themselves. Terrific killing took place, especially of deer and antelopes which had been protected to provide sport for the rulers and their guests. In some states animals such as deer were slaughtered just for the sake of slaughter, as only a tiny percentage of the animals could be eaten.

Five years elapsed before some sort of wild life conservation was started. In 1952 the centralised Indian Board for Wild Life was inaugurated to take charge of wild life and from that date onwards Wild Life Boards were created in the various states. Although wild life in most other countries is under the jurisdiction of a special department, in India nearly all the wild life and all the sanctuaries are either inside or at least under the control of the Forest Departments of the states, which have special Wild Life Divisions and separate staff for wild life. Looking back now it seems that this system has not worked very well, but it is probably the only system which is at all practicable in India. It is up to governments and people to make it work for there is no reason why it should not function if everybody cooperates.

Mention has been made of the fact that as recently as fifty years ago, wild life was plentiful almost everywhere in India. And yet some species are now very rare while others are on the verge of extinction. There is very little information at the present time as to the exact numbers of the more important species, except that we know that the population of the wild asses in the Little Rann of Kutch is about 860, and that the population of the Indian rhino in West Bengal and Assam (and Nepal) is about 740.

Although censuses were not attempted in the past, a study of old *shikar* books and the memory of older people who are interested in sport and natural history, indicates that only about one-tenth of the wild animal population of fifty years ago exists today in respect of tigers, leopards, all species of deer and antelope, and most other larger mammals. This is an unreliable method of calculation, but it is better than nothing in the absence of alternatives. It should be borne in mind, however, that it is essential to conduct proper censuses of the more valuable and rare mammals and birds.

The Royal Bengal tiger must surely be the best known of all

196

India's animals. Half a century ago it is estimated that there were some 40,000 tigers, and even in the early 1950s they were said to be plentiful. It thus comes as a surprise to most people to learn that the tiger is now threatened with extinction in the wild, with the experts estimating the population at fewer than 3,000. Destruction of the forests, over-hunting, and use of agricultural chemicals to poison tigers near villages have brought this magnificent creature to this pass. Its future lies in sanctuaries, but they have to be large enough for a healthy breeding population, and well stocked with the deer, pig and other creatures on which the tiger preys.

Among the small animals, lack of sightings of the Pygmy hog, which is less than a foot high, or the Hispid hare, led to the belief that they were extinct, which was attributed to destruction of their habitat along the base of the Himalayas. Then, in 1971, a party of tourists saw and photographed Pygmy hogs in the Manas sanctuary in Assam, and soon afterwards a tea garden manager was able to obtain a dozen specimens after he heard that local people were catching them where a grass fire had broken out. The Hispid, or harsh-furred hare, which has unusually short ears, was caught at the same time, but unfortunately just one and so there is no opportunity to try captive breeding.

The reasons for the failure of the cheetah in the battle for survival appear to be firstly that the cheetah, and its chief prey the blackbuck, inhabited open places needed for agricultural cultivation and grazing rounds for domestic livestock. With the increase in human population and domestic livestock population, their habitat became more disturbed and the number of blackbuck was reduced. The cheetah no longer stood a chance of survival. Secondly, cheetahs have never bred in captivity in India, and very seldom in other parts of the world, and consequently there was always a continuous demand for wild cheetahs for zoos and for training as hunters of blackbuck.

Probably the only animal in India which has actually increased in numbers during the past fifty years is the great Indian one-horned rhinoceros. At the turn of the century there were many rhino in Nepal where, with tigers, they were regarded as 'royal game' and strictly protected for exclusive hunting by the rajas and their guests. But in the Jaldapara area of northern Bengal and in the Kaziranga area of Assam they were nearly exterminated by poachers. The Kaziranga population was believed to be only about one dozen then: now it is 400. Great vigilance has to be maintained against unscrupulous poachers who are after the horn – which is mistakenly but persistently believed to be an aphrodisiac, especially in the Far East where most of the poached horns are known to find their way. Even in this modern technological age the market price of rhino horn has considerably increased to about £250 per pound, and not long ago a forest guard on duty in the sanctuary was murdered by poachers intent on acquiring more rhino horn to satisfy this idiotic demand. The Indian rhino is rarer than the two species of African rhino, but not so rare as the two-horned Sumatran rhino and the lesser one-horned Javan rhino. It used to be found as far west as Kashmir, but now only exists in north-east India and Nepal. It is solitary in its habits and if two are found together they must be either a courting couple or else a mother and child. In spite of being thickly covered with 'armour plating' and rather unintelligent, it can jump and gallop and easily outstrip a human being or an elephant and rider.

One species which may have declined in numbers but about which there is no need for concern is the Indian elephant. Being very prolific, the elephant has survived the inroads made by ivory hunters and the reduction of its numbers by elephant-control licence holders who are encouraged to limit the numbers of this large-footed and large-mouthed beast when crops are being raided and damaged. Their numbers appear to be somewhat static now, at the level of about 6,000 in the whole of India. Control of wild elephants, by trapping operations, by *khedda* (stockade) or by *mela shikar* (noosing in the forest), or in pits, is becoming less and less profitable and therefore less popular with elephant-catching companies because of the diminishing demand for trained elephants in this modern mechanical world.

From the point of view of wild life, however, there is nowadays an interesting function for trained elephants for which they will always be in demand. Trained elephants continue to be required as mounts to take visitors round wild life sanctuaries – especially those where the grass is high or the forest thick. Kaziranga Sanctuary in Assam, for example, usually maintains about sixteen riding elephants; and Jaldapara Sanctuary in West Bengal, Corbett Park in Uttar Pradesh, Kanha Park in Madhya Pradesh, Bandipur Sanctuary in Mysore and Mudumalai Sanctuary in Madras maintain a smaller number of riding elephants. In thick or boggy country riding a well-trained elephant is not only the best way to travel and see the wild life but is also a joy in itself for visitors, especially those from abroad to whom it is a novel and exciting experience.

Reference has already been made to the Indian lion and how its numbers became seriously reduced during the last century. It finally became extinct everywhere except in the Gir Forest. At the beginning of the century there were believed to have been only about 100 lions left – the lowest population ever of this noble beast of prey. A rough census gave the population as 240 in 1950, 290 in 1955 and 280 in 1963; but owing to the fact that a number of lions are known to have been poisoned by the angry owners of domestic livestock who have had animals killed by lions, the population may well be nearer the 150 or 200 mark. It is to be hoped, however, that the recent introduction of a system for paying compensation for domestic animals killed will put a stop to this practice. The recent upgrading of the Gir Forest into a sanctuary should also be in the interests of the lions. The Indian lion very closely resembles the African one, except that the manes of the males are slightly smaller. It is an older inhabitant of India than the tiger and, contrary to popular belief, the tiger is not responsible in any way for the decline of the lion. As was pointed out earlier, it was excessive killing by 'sportsmen' that reduced the numbers of the lion so drastically, for the lion is a diurnal animal and more bold than the tiger, thus making a better target. There is no admixture of African blood in the lions of India, although three pairs of African lions were indeed brought to the former Gwalior state about the year 1916, but these were released and later shot as far away from the Gir as Shivpuri (in Madhya Pradesh) and could not possibly have reached the Gir Forest. The food of the lions consists both of the other wild animals of the area such as nilgai, sambar, chital and wild

199

pig, and also of the local domestic livestock. Occasionally the lions are fed with dead buffalo baits to provide a spectacle to visitors; and this kind of 'lion show' is proving to be a great tourist attraction.

Another instance of a ruler of an erstwhile princely state protecting an animal species is to be found in Kashmir. The Kashmir stag, a subspecies of red deer, used to be 'royal game' in that state, and there may have been as many as 5,000 of them fifty years ago, and about 2,000 left in 1947. In the following years the unsettled conditions in this lovely part of the world led to a further reduction of this fine deer to about 300. Now there may be about 200–300 left in Dachigam Sanctuary, one of the most beautiful places in the world. These magnificent deer spend the winter in Lower Dachigam at about 5,500 feet, and in the spring they migrate upwards into the mountains in Upper Dachigam and elsewhere to about 16,000 feet, returning again in late autumn. Because of their migrations these deer are not easy to see, except in the depth of winter, in January and February, when snow lies thick on the ground.

While on the subject of high elevation wild life, the present status of the snow leopard, the wild goats (ibex, markhor and Himalayan tahr), wild sheep (shapu or urial, nayan, ovis ammon and bharal), goat antelopes (takin, serow and goral) and musk deer is not known because of frontier security precautions. It is feared that many of these interesting species may have been decimated by military personnel both on the Indian side of the border and on the Chinese side. An encouraging item of news, however, came from the Kazinag area of Kashmir. Here, close to the ceasefire line, the markhor is believed to have increased due to the fact that there have been no shots fired for fear of starting off a new war. [Unfortunately shots *have* been fired since this was written. Eds.]

Not very far from the Indian lions of the Gir Sanctuary are the wild asses of the Little Rann of Kutch, also in Gujarat State. These asses used to be reported 'in thousands', but no census or even detailed estimate had ever been made till 1962 when it was found that only about 860 existed. Another detailed estimate made in 1966 gave a similar figure. These handsome asses are larger than the local donkeys, which are

200

believed to be descended not from them but from Egyptian asses. They are a bright sandy colour with mane and dorsal stripe of dark chestnut and white lower parts. Occasionally, so-called 'sportsmen' go there and run down some of these asses in jeeps, and some are known to have died of South African Horse Sickness and other diseases in 1960. But their numbers seem to be more or less static. The local villagers do not molest them, they are vegetarians and have a high respect for the sanctity of life. Even when locusts came and devoured their crops they wanted only to drive them away, not to kill them. In order to get photographs of these asses one has to follow them on to the Rann in a jeep doing up to 34 miles per hour.

Another animal which has become very rare and is on the danger list, and therefore fully protected, is the graceful brow-antlered deer of Manipur in north-east India. In 1951 it was officially stated that it was extinct; but a few years later it was seen in a very swampy area of tall grasses and reeds. In fact this place is a floating swamp, in other words a thick mat of humus and dead vegetation which actually floats on the water of the Logtak Lake. Like an iceberg, one fifth of this mat is above and four fifths are below the surface of the water. And on the mat of humus grow reeds and grasses up to fifteen feet in height.

These brow-antlered deer, so-called because of the con-tinuous graceful curve of the brow tines and the beams of its antlers, have lived in this place for thousands of years and have developed slightly splayed-out hooves and their pasterns are hairless and horny so that they can walk with them bent down and not sink through the floating humus. About 100 of these rare deer are believed to exist in their ten-square-mile sanctuary called Keibul Lamjao, and cannot normally be seen by visitors. But two small herds of them are thriving in the zoological gardens of New Delhi and Calcutta and these are on view.

Much of India's wild life is peculiar to this country or at least to the subcontinent, and is not found in other parts of the world. Mention has already been made of the great Indian rhino which is only found in India (with some in Nepal and perhaps a very few in the extreme north of Burma), and of the

Asiatic lion which is only found in the Gir Sanctuary, having become extinct in all other parts of Asia.

Of antelopes, the nilgai and four-horned antelope are almost only found in India (with a few in Pakistan). Of deer the Indian swamp deer is found almost only in India (with some in Nepal), and the most beautiful chital or spotted deer is found almost only in India (with some in Pakistan, Nepal and Ceylon). The blackbuck is not found elsewhere than in India and Pakistan. Something has already been said about how the Indian cheetah became extinct because its prey the blackbuck was becoming scarcer: well, blackbuck have steadily decreased. Where thousands previously existed, only tens or even fewer are found now. In some places they are extinct. Being far-ranging creatures of the wide open spaces, they are difficult to protect in sanctuaries. Fortunately a few private preserves still exist, and these may be the only hope for the blackbuck's survival.

The wild buffalo is also found only in parts of India such as Assam and Madhya Pradesh, and Kaziranga Sanctuary in Assam is the best place for seeing these. The Indian 'bison' must be mentioned here, for it is not a real bison at all but a wild ox – in fact it is the largest of all the world's bovines and is a truly magnificent creature. It is also found in Burma and Malaya. In India the best place to see fine creatures is Kanha Park in Madhya Pradesh – where they have actually increased slightly in numbers.

With game preservation in the old game sanctuaries of the British and in the game preserves of the princes evolving into modern wild life conservation in sanctuaries and national parks, the emphasis nowadays should be on observing and photographing wild life rather than on killing. Photo *shikar* or 'shooting' with a camera is in fact gradually taking over from shooting with a gun, and is far more exciting and rewarding. Even the *shikar* agencies are including photographic safaris within their scope, because obviously there will not be enough tigers and other such animals for big game shooting very soon.

Not only has killing by poachers and sportsmen reduced the numbers of wild life throughout India, but also a terrific threat has arisen through loss of habitat due to ever-increasing numbers of human beings and their livestock. Everywhere the forests and other wild places are receding in the face of extending cultivation and grazing grounds. Even the reserved forests, sanctuaries and national parks are not altogether immune from trespass and grazing. Due to the pressures from increased human and livestock populations, and due to operations connected with exploitation of forest produce, the reserved forests throughout India have become sadly depleted of their wild life in recent years. This means that nearly all the larger mammals of India (not birds, which are more able to fend for themselves in open areas) are to be found in the national parks and sanctuaries and depend on their protection in these places for their ultimate survival.

National parks in India are only a few in number. The first to be created was the Hailey National Park in Uttar Pradesh in 1935, now called Corbett National Park after the famous hunter of man-eaters. This is one of the places where there is the greatest chance of seeing a tiger. Since Independence several more national parks have come into being; the Kanha and Shivpuri National Parks in Madhya Pradesh, the Taroba National Park in Maharashtra and the Hazaribagh National Park in Bihar. In addition several of the best sanctuaries have been approved by the Indian Board for Wild Life for up-grading into national parks by the states in which they are situated. Of these the most notable are Gir Sanctuary in Gujarat, Kaziranga and Manas Sanctuaries in Assam, Periyar Sanctuary in Kerala, Mudumalai Sanctuary in Madras and Bandipur Sanctuary in Mysore.

The question is often asked as to which is the best of India's wild life places. Well, it is not easy to pinpoint any park or sanctuary in India and say it is the best one. For there are about seven or eight places which are all equally outstanding in their own special way with their own typical and representative fauna and flora, and could well hold their own in competition with the famous show places of Africa and North America.

The Gir Sanctuary in Gujarat can provide wonderful views of lions' quarters in the typical dry thorn forest of that region; while across in the north-east of India Kaziranga in Assam and to a lesser extent Jaldapara in West Bengal provide lush green grassy swamps with rhino, wild elephant, wild buffalo and several species of deer; and midway between these two places is the Manas Sanctuary of Assam and Bhutan with its spectacular mountain and river scenery and wild life. Again in Central India, a little more difficult of access but very well worth the effort of getting there, is the wonderful Kanha National Park of Madhya Pradesh with its beautiful grassy tree-dotted meadows in which swamp deer, chital, blackbuck and gaur roam and on to which a tiger sometimes strays.

In South India it is very easy to see Handipur Sanctuary in Mysore and Mudumalai Sanctuary in Madras on the same visit, for the two are continuous, and Rangathittoo Bird Sanctuary can be seen in just a few hours while staying in Mysore city. Further south are the scenic lake and wild life of Periyar Sanctuary in Kerala, which should certainly not be missed by anyone travelling in South India. Vadanthangal Bird Sanctuary near Madras can also be visited if it is the right time of the year – November to February. Easily accessible from Delhi are the Keoladeo Ghana Bird Sanctuary of Bharatpur, and the Sariska Sanctuary which is not far from Jaipur city. In this latter place there is a fair chance of glimpsing a tiger – probably the most sought after wild creature in the world.

India has always been known as a country of monkeys and snakes. The visitor will almost certainly see the common Rhesus monkey, which has adapted itself to living in the towns and villages including the city of Delhi. It is still plentiful, although somewhat diminished in numbers through trapping for export for medical research and production of vaccines. The graceful long-tailed Langur, with its black face, is more retiring, but often frequents temple areas and can be seen along country roads. Unhappily some monkey species are threatened with extinction through over-hunting for their beautiful skins and meat. This is especially true of the Nilgiri Langur and the Lion-tailed Macaque, which are found in southern India.

As for snakes, you could live for years without seeing one, except those constantly displayed by snake-charmers. They are there, however, but most have a retiring disposition and keep out of the way of humans. Thousands of people are bitten by snakes every year, usually because they step on them with bare feet at night – snakes don't like being trodden c n. But only four out of the many common varieties of snake in India are poisonous, and a large number of 'snake doctors' make a good living by removing the fears of people bitten by snakes – which can prove mortal in themselves – through some abracadabra. A snake farm on the outskirts of Madras, established with the aid of the World Wildlife Fund, displays a wide range of species and explains the beneficial role of snakes as consumers of the rodents which eat a great deal of India's precious food grains.

These sanctuaries and national parks of India, with the wild life they contain, provide a very great tourist attraction for visitors from all over the world. India can indeed rival Africa in her richness and variety of wild life, and can surpass many other parts of the world with her magnificent mountain, forest and river scenery. And as there are such diversities of climate (from the driest to the wettest in the world), and of elevation (from sea level to the high Himalayas), there are at least some wild places with unique wild life which can be visited and enjoyed by tourists from abroad and visitors from within India at any time of the year.

Birds in India

Peter Jackson

Kites and vultures wheel and soar in the sky; sleek mynahs strut on the lawns where black and white striped hoopoes probe for grubs; screaming green parakeets streak overhead; and the merry warble of the bulbuls fills the air. Birds impress themselves upon the visitor to India as perhaps nowhere else in the world.

In the heart of the cities, in barren deserts, in hot humid forests, and in the arctic climates of the high Himalayas the birds are active. India has about 1,200 distinct species, and if you start to count sub-species you can bring the total up to about 2,000 varieties from the glamorous peacock, which has been adopted as India's national bird, to the dull, ugly-looking vultures, and from tall stately cranes, to tiny jewel-like sunbirds and warblers.

The capital city of Delhi offers as good an introduction to India's birds as anywhere. It lies near the eastern edge of the Rajasthan desert, while to the east of it is the 1,000-mile long Ganges valley running to Calcutta. The River Jumna, a tributary of the Ganges, runs north-south by Delhi and is a favoured migration route for swarms of birds which fly in to winter on the Indian plains, and then move north in spring to breed in Central Asia. Over 400 species have been recorded within a 30-mile radius of Delhi, and on a winter's day it is not difficult to see over 100 different kinds of birds.

From October greylag and bar-headed geese can be heard honking as they fly over the city at night. They settle with thousands of duck on the marshes, where the shores are full of sandpipers, plovers, and other waders.

Siberian chiffchaffs flit through the trees, along with lesser whitethroats, and a host of tiny warblers, which appear much too frail to have undertaken the adventurous flight over the

206

world's highest mountains.

These migrants have come to join the residents who are able to stand the blazing heat of summer, and in fact have found that this is the best time of the year to breed.

On crisp winter mornings the duck hunters are out by dawn, crouching in cover by the *jheels*, as the marshes are called, awaiting the flights of pintail, shoveller, teal, gadwall, and pochard which swish out of the sky. Usually they are hoping that a skein of geese will pass low over them. Nowadays the bags seldom equal those of the old days. In 1938 the British Viceroy, Lord Linlithgow, went on a duck shoot at Bharatpur, about 120 miles south of Delhi, and with 41 guns in action 4,273 birds were brought down. The Bharatpur marsh has been declared a sanctuary and is one of the most spectacular in the world.

While the hunters are concentrating on their duck, let us look around. That mass of white birds consists of several hundred spoonbills moving together and sweeping their slightly open bills from side to side, snapping them shut when they find food. With them are white ibises with bare black heads and long curved bills.

The white birds scattered along the edge of the marsh are egrets, which were once much persecuted for their delicate breeding plumes which were used to decorate ladies' hats. Egret farms were established in some areas to maintain the supply, but happily the fashion has died out and the egrets are safe.

A loud trumpeting rings out and draws attention to the tall grey sarus crane. These stately birds are much revered in India and a pair may be found by many village ponds. Legend has it that they mate for life and are so devoted that if one dies the other succumbs from a broken heart. It is true one never sees a lone sarus crane, but experts tell us that a lost mate is soon replaced.

Waders are scurrying along the shores of the marsh, sometimes exploding into the air to wheel in tight flocks, their white undersides flashing in the sun, before they settle again to their feeding.

Contrasting with the nervous activity of the waders is the controlled stealth of the herons, standing alone and peering

207

intently at the water before making a sudden dart to grab a hapless fish. Grey and purple herons are obvious, but there are also small pond herons, or paddy birds, whose plumage makes them inconspicuous until they fly with a flash of white wings.

Rosy pelicans surge through the deeper waters like a fleet of galleons in full sail, while flamingoes stalk along on spindly legs and sieve the mud for food through their inverted bills. An osprey glides overhead and plunges suddenly to catch a fish, while eagles of several kinds soar on almost motionless wings, or sit quietly on the trees. Pied kingfishers hover with their black bills aimed menacingly downwards.

The calls of all these birds mingle, but the scream, "did-he-do-it", of the red wattled lapwing rings out above all. It is the bane of the hunter trying to stalk his prey, for the lapwings always spot an intruder and advertise his presence to all and sundry. But these lapwings are not especially shy birds. Indeed they nest in gardens and on the flat roofs of houses in towns and cities. One pair built a nest of over six hundred chips of cement on the roof of the Defence Ministry in Delhi. It was not an ideal nest in the summer temperatures of well over 100 degrees Fahrenheit and, although the eggs began to hatch, the young ones died before they emerged from the shells.

As the winter visitors depart the resident birds start nesting. Every household starts its annual battle with the sparrows who take advantage of open doors and windows to fly in and build nests on bookshelves or any other niche considered suitable – by the birds, that is.

Purple sunbirds build intricately woven hanging nests and tailor birds sew leaves together only a few feet from where people walk to and fro. Despite the building activity the nest is frequently overlooked.

Squawking and quarrelling, the mynahs draw attention to themselves as they struggle for possession of suitable holes for nesting. These holes are always a source of contention in the bird community. Parakeets, owls, woodpeckers and mynahs dispute ownership. I once saw a mynah get a nasty shock when entering a hole. An irate woodpecker stabbed it in the face and the mynah fell off in confusion.

While India has abundant bird life, there are some species

which have become extinct in recent years, and others which are threatened. Birdwatchers still cherish the faint hope of seeing the pink-headed duck. It was never very common and the last report of one was in Bihar in 1935.

In the extreme north-eastern areas of Assam and Manipur there are still some white-winged wood duck, but they are becoming very scarce and may become extinct.

The great Indian bustard stands four feet high and is one of the most magnificent birds in India. But pressure of population and development of agriculture has greatly reduced its habitat. Its flesh is highly esteemed and there has been continual poaching. The jeep has played its part in the slaughter for, although the bustard is exceptionally wary and difficult to approach on foot, it is not worried by noisy vehicles and the poachers get them easily. The government has declared the great Indian bustard as totally protected, but enforcement of protection is extremely difficult.

No fears of extinction arise over the Indian house crow. It thrives along with human kind. Mark Twain was taken with them when he visited India, despite having his breakfast stolen by them. He wrote: 'Their number is beyond estimate, and their noise in proportion. I suppose they cost the country more than the government does; yet that is not a light matter. Still, they pay; their company pays; it would sadden the land to take their cheerful voice out of it.' The humorist went on to declare that the crow had accumulated 'all damnable traits' with the result that 'he does not know what care is, he does not know what sorrow is, he does not know what remorse is, his life is one long thundering ecstasy of happiness, and he will go to his death untroubled.'

Travelling around India you can easily notice the different races of the crow produced by environment. In the north-west the crows have a light grey mantle, but southwards and eastwards the mantle gets darker and darker, until in the south it is a dirty browny grey which barely shows against the rest of the black plumage.

Alert and active as the crows are, they are still outwitted by the koel, a member of the cuckoo family which relies on the crows to raise its young. The sight of the sleek black koel or its speckled wife drives the crows to paroxysms of fury. They launch furious attacks on the koels, but still the female manages to deposit her eggs in the crows' nests and leave them to take parental responsibility.

The crescendo broken shriek of the koel is one of the familiar sounds of the Indian summer. Similar to it, but more harsh, is the call of the common hawk cuckoo, which has been rendered as 'it's hot, it's hot, it's hot, it's hot. . . . brain fever . . . brain fever'. In fact more people know it as the brainfever bird than as a cuckoo.

The crimson-breasted barbet has the sobriquet 'coppersmith' because its monotonous call sounds just like the artisan tapping away at the metal.

There are some gay girls among Indian birds, notably the painted snipe, which is not a snipe at all, but one of the rails. The females are much more brightly coloured than the males, and the reason becomes obvious when their family roles are looked into. The females fight for possession of the males, but having laid the eggs, they go off and look for another mate. Because he is left to incubate the eggs the male painted snipe has assumed the secretive colouring usually found in females.

This reversal of family roles has also been found recently to be true of the pheasant-tailed jacana, although here the plumage of males and females is identical – chocolate brown, with white head and wings, and a yellow nape, and a long, curving, pheasant tail in breeding plumage. The males mark out the breeding territory on lakes and marshes and the female helps defend it while she is his mate. But when the eggs are laid she goes off to find another male.

The jacana has some other interesting habits. When disturbed it will build another nest and move the eggs there. I was once photographing a jacana on its nest in Kashmir, and the bird was continually pulling at the floating vegetation a few yards away. I thought it was feeding, but suddenly it came to the nest, rolled one of the eggs into the water and began to back away, tapping it along with its bill. It took the egg to a new nest it had been making when it appeared to be feeding. One by one it moved all four eggs there and then settled down. The eggs were probably nearly ready to hatch so that they floated, but jacanas have been seen carrying eggs pressed between throat and breast, or rolling them over vegetation when they were probably at an earlier stage of development.

There is a different order of things among the weaver birds, whose retort-shaped nests hanging in groups in the trees are a prominent feature of the countryside. Drab and sparrow-like in winter plumage, the males develop glorious yellow caps and bibs in summer. In noisy parties they peel strips from grass and weave their intricate nests, suspended from small branches. The nests have an egg chamber, and alongside it is a vertical tube up which the parent birds enter.

When the nests have been built the females arrive. They pay no attention to the males, but carefully inspect the nests. Having found one they approve they move in, mate with the male who built it and settle down to brood the eggs. The males, however, move off to build more nests and await the arrival of more females. In a season they may build three or four nests and have that number of wives.

The weavers are a heavy burden to the farmer for they are great grain eaters, along with the parakeets and the peafowl. During the season small boys are kept busy banging tins and shouting and waving in what is often a fruitless effort to keep the birds away from the crops.

The forested hills of India have an extremely varied and colourful bird population. Seeing the birds is not always easy, and you can sit in the forest and wonder if there are any birds there at all. Then suddenly your particular bit of forest is alive with many different species – tits, warblers, babblers, thrushes and others. They band together in mixed hunting parties, in which each species benefits from the feeding activities of the others in uncovering insects or seeds.

Resounding across the valleys is the piercing 'phew, phew' of the great barbet, a relative of the coppersmith of the plains. Several barbets sometimes seem to be trying to outdo each other in noise and duration. Blue magpies, with long streamer tails, play follow-my-leader through the trees. In the glens the liquid call of the whistling thrush, which is so human that the bird is sometimes called the 'whistling schoolboy', can be heard. White-capped redstarts hop from boulder to boulder in the stream beds.

The dull brown dipper plunges into the raging torrent, and seems sure to be swept away. But he pops up again, his skill at

211

manipulating his wings under water to keep a footing having ensured his safety. Even if you know the trick, it still seems miraculous that the dipper can survive in such waters.

A cackling chorus in the forest comes from a party of laughing thrushes on their daily hunt for food. They are an interesting family, for although they are poor flyers and do not migrate, they are found on the Himalayan slopes and then again over 1,000 miles away on the southern ranges of India, but not in between. Several other species are found in similar circumstances, as if the hills were islands in a sea of plains.

According to India's renowned ornithologist, Dr Salim Ali, it is believed that the laughing thrushes may once have been distributed over the whole of India. Outside the hill areas they may have become extinct because other species were better adapted. Or the elevated country between the Himalayas and the southern hills may have worn away leaving isolated relic communities. It is interesting to find that the same vegetation is to be found wherever there are laughing thrushes.

Some other hill species do migrate between the Himalayas and other ranges. The woodcock, the blue chat, and the pied ground thrush are examples, and since they are seldom met with between the Himalayas and the Nilgiris, and related hills in peninsular India, it must be assumed that they make hops of nearly 1,500 miles non-stop.

The Himalayas rise from a few hundred feet to Everest's 29,028 feet in a matter of only about 60 miles, and they have a range of climates from humid tropical to alpine and arid in which there are tremendous variations in vegetation. In this compressed sample of world climate many species of birds have become altitudinal migrants, frequenting the lower ranges in winter, and even visiting the plains, and then moving to the high elevations to breed in spring and summer. Tits, warblers and chats are among the species which do this.

The crags are the nesting places of the huge lammergeirs, which are also called 'bearded vultures' because of the neat goatee beards they sport. They ride majestically on the air currents over the hillsides and have the peculiar habit of dropping bones from a height to break them open and expose the marrow, which is the bird's special delight.

One of the most extraordinary bird 'events' in India is in

the North Cachar hills of the north-east, near Haflong. About September there is an appalling massacre of birds by the local people, with the birds appearing to offer themselves for sacrifice. It seems that around the turn of the century a party of local villagers were out one misty, moonless night looking for a buffalo feared to have been taken by a tiger. They carried flares, and suddenly hundreds of birds descended on them and settled on their arms and shoulders.

The birds were taken as a god-sent gift to compensate for the loss of the buffalo, and they were killed and eaten. Later it was found that at the same season and in similar conditions the phenomenon occurred every year, but only at this one particular spot. And so the people regularly put out bright lamps and then hide behind a screen. Large numbers of birds come flying into the lights and are beaten to death.

Ornithologists assumed that because September is the time of the southward autumnal migration the birds must be migrants. But when Dr Salim Ali and the late Mr E.P. Gee collected remains of the birds, they were found to have been resident birds such as the white egret, kalij pheasant, green pigeon, white-breasted kingfisher, laughing thrush, and emerald dove. Despite official efforts to stop it the slaughter is believed to continue each year.

During the monsoon India offers bird spectacles scarcely equalled anywhere else in the world. Huge numbers of water birds gather in vast nesting colonies, nest upon nest, until it seems the trees ought to collapse. These colonies may be found in many parts of India, sometimes alongside, and even partly in, villages and small towns. The most famous is at Bharatpur, and is easily reached from Delhi.

Big yellow-billed painted storks like the tops of the trees, where they stand motionless for hours, sometimes with their wings partly or fully spread to protect their young from the blazing sun. Lower in the trees are spoonbills, cormorants, and egrets. White ibises pack their nests so closely together that when the young are becoming fledged the tree is hidden by a seething mass of white.

Darters, whose long slim necks which stick up out of the water as they swim have given them the name of 'snakebirds', have a dark brown plumage, but their young are pure white.

If you put your hand near a nest of young darters it is indeed like putting it near a nest of striking snakes.

Open-billed storks nest in large numbers at Bharatpur. They have peculiar pincer-shaped bills with a gap in the middle through which you can see daylight. This is probably an adaptation to enable them to retrieve the large molluscs they like for food from the mud of the marsh.

The mind boggles at the thought of the amount of food necessary to sustain this vast concourse of birds. One rough estimate puts the requirements of several thousand painted storks alone, occupying just one square mile of the marsh, at between three-and-a-half and five tons a day. Given a normal nesting period of 30 to 40 days this means that 105 to 200 tons of fish are consumed by just one small part of the bird community. Other birds, such as cormorants, eat far more fish than the storks.

The sound of the heronry is one continuous roar, rising to a crescendo like that of an excited football crowd as an eagle or other predator glides overhead in leisurely fashion looking for a meal. The parents do not defend their young. Usually they just take off and flap around. The eagle makes its selection, dives, lifts a young bird and flies away to tear it to pieces and feed. The heronry immediately relaxes and life returns to normal. Despite the toll exacted by the eagles, they can often be seen sitting right in the middle of the colony without causing any disturbance. It appears that the birds know when they are hunting.

With such a wealth of bird life around, it was inevitable that birds should come to play their part in the mythology of the peoples of India. They appear with many of the Hindu gods and goddesses as their vehicles. The peacock is the mount of Saraswati, the Goddess of Wisdom. Brahma, central figure of the Hindu Trinity, rides on a swan, while his fellow Vishnu has the white and chestnut Brahminy kite as his vehicle.

Some Hindu religious books refer to birds as exemplars of various qualities. The heron stands for hypocrisy and treachery; the cock for discrimination; the thrush for secretiveness; and the sarus crane for caution. The owl, which Westerners often use as a symbol of wisdom, usually means just the opposite in India. To call someone *Ullu* means he is a fool.

Birds play an important part in the decorative arts. The peacock with its tail spread is carved in wood and stone, and marked out with precious stones. It was used for the fabulous Peacock Throne of the Mughal emperors, which was taken away by Persian invaders and broken up. But even today the throne of the Shah of Iran is called the Peacock Throne.

In Kashmir the little blue Siberian kingfisher, a species of the common kingfisher found throughout Eurasia, is used a lot in intricate paintings on *papier mâché* and on carved screens. This is not surprising, for a visitor to the Vale of Kashmir is immediately taken by the numerous kingfishers which sit on the houseboats and plop into the water to catch a fish, or race across the lakes and waterways emitting a high-pitched squeak.

The paradise flycatcher, the male pure white with an indigo head and crest, and long streamer tail feathers, appears frequently in decoration. Kashmiris call the bird the 'cotton-flake' – a delightfully descriptive name for a bird which does seem to have the lightness of a cottonflake as it flits through the trees.

One of the early birdwatchers in India, who recorded his observations in his own hand until 1622, was Emperor Jehangir. He commissioned his court painters to illustrate them for his diary, *Jahangir-nama*. However, most of the scientific writing about Indian birds began with the coming of the British. Administrators, soldiers and policemen became interested in the bird life around them and they often had ample time to observe and make notes. Many were keen on shooting, and some of the best of the early books concentrate on game birds. They contain excellent illustrations.

The Bombay Natural History Society is the premier learned body dealing with ornithology in India, and its Journal contains important notes and articles. Associated with it is the Birdwatchers' Field Club of India, which publishes a monthly Newsletter bringing together the notes and experiences of people from all over the country.

For a select list of books an Indian Birds, the reader should turn to R.E. Hawkins's chapter on 'Books About India'.

Books About India

R.E. Hawkins

For daily use acquire a Murray. It doesn't much matter which edition, for all have their charm. The first of 1859, for instance, remarks on p. ii:

> It happens that many of the most interesting Indian localities are situated among thick jungles, loaded with noxious vapours and abounding with dangerous reptiles and wild beasts. Thus the caves of Salsette can never be securely examined by the traveller; and no one should explore the ruins of Mandu unless fully equipped for a tiger hunt.

My edition, the twentieth of 1965, describes the approach to the cave temples of Kanheri more prosaically:

> Train or car to Borivli station, 19m. from Bombay (good clean waiting room); thence 5m. to the caves by a rough country road, on which carts ply and which runs to within 1m. of them.

In fact there is today a good motor road all the way, along which buses ply. On how to get about and where to stay more recent guide books are more reliable, Fodor's *India 1972* (1971) or Willcock & Aaron's *India in $5 & $10 a Day* (1970) are generally satisfactory in this respect but must be regarded as supplements, not substitutes. Once at your destination, however, Murray provides mines of information. The title of the twentieth edition is *A Handbook for Travellers in India, Pakistan, Burma and Ceylon*. While Fodor contains two pieces on the Kanheri caves they are partly contradictory and partly repetitive and are scattered over two pages (345 and 360) without cross-references. *India on $5 and $10 a Day* merely mentions the caves, but Murray devotes a full page to them. All these books are unsatisfactory in their maps. Murray's

216

are more numerous and accurate but are out-of-date. Plans such as those of Mandu, Cochin and Bhubaneshwar in Fodor are useful but scarce. Incidentally it is advisable to acquire Murray before arriving, for copies sold in India usually lack the map in the pocket at the end.

In the first edition Murray gives a two-page list of books 'which may be perused by the traveller previous to starting'. The Select List in the twentieth edition is much shorter, pointing out that 'the Literature on these countries is enormous' and referring readers to the body of the Handbook for books on special subjects or places. In this chapter I shall have to be even more selective.

It is in Rudyard Kipling's *Jungle Books* that many people first visit India, and a wonderful place it is. Hathi the elephant, Shere Khan the tiger, Baloo the bear, Bagheera the leopard, Kaa the rock-python, Akela the lone wolf and dominating all Mowgli the man-cub, suckled by Akela's mate – these names are familiar throughout Europe and America. In the stories themselves many of Kipling's predilections are implicit: his admiration of courage, action and discipline, his awareness of universal and inevitable law, the complementary natures and duties of the jungle animals, and man's loneliness and complexity, both beast and man. Though universal in their scope and appeal, the fact that the background of the stories is the Indian jungle and Indian village life leads the visitor to associate such qualities with India, besides introducing him to the sunlit and moonlit jungles of Madhya Pradesh which, though shrunken in extent and riven by roads, still survive and are one of India's chief attractions to megalopolis dwellers.

Rikki-tiki-tavi the mongoose is a familiar friend too, and if Kipling's father had not said that he never possessed a gun (*Beast and Man in India*, p. 57) I should have thought it was in the garden of the School of Art bungalow in Bombay, where Rudyard was born in 1865, that Rikki-tiki-tavi danced around and feinted with the cobra until her head was blown off.

When he was six Rudyard was packed off to England and he only spent another seven years in India, (neglecting the short visit in 1891) from the age of 18 to 24, working as a journalist first on the *Pioneer*, Lahore – 'the greatest Indian daily paper' he proudly called it in *The Miracle of Purun Bhagat* – and then for a couple of years on the *Leader* at Allahabad. While he was still in India he published *Departmental Ditties* (a measure of their popularity, and of Kipling's commercial flair, is offered by the fact that Kipling bought back his copyright in this collection of fifty poems for £2,000 in 1899), *Plain Tales from the Hills* and *Soldiers Three*, 'setting forth certain passages in the Lives and Adventures of Privates Terence Mulvaney, Stanley Ortheris, and John Learoyd'. *The Naulahka*, written in collaboration with Wolcott Balestier, the American whose sister Kipling later married, appeared in 1892, soon after its author had left India, *Many Inventions* in 1893, *The Jungle Book* in 1894, the *Second Jungle Book* in 1895 and *Kim* in 1901. Kipling's Simla society, his carriers of the 'white man's burden' (1899, already passed to the USA,) and his British soldiers in India have all gone, but the background to his stories – the Himalayas, the jungles of central India, the bazaars of Lahore and Lucknow, the ancient cities of Rajasthan: Amber, Chitor and Jaipur, the ceaseless variety of the Grand Trunk Road – all these are still to be seen. Kipling's view of India and its peoples may be a superficial one, and it is certainly a view of a very small part of the whole, not much more than Lahore, Simla and Rajasthan, but it is photographically accurate for both people and places.

The Nobel prize in Literature was awarded to Kipling in 1907, when his reputation rested largely on his books about India, and in 1913 the same prize went to Rabindranath Tagore, the Bengali poet. The prize must have been awarded on the strength of his work in Bengali, for very little had at that time been translated into other languages. In his introduction to *Gitanjali*, written in 1912, W.B. Yeats said: 'these prose translations from Rabindranath Tagore have stirred my blood as nothing has for years', and Edward Thompson, who knew Bengali, says: 'he wrote, with a skill and virtuosity hardly ever equalled, in a tongue which is among the half-dozen most expressive and beautiful languages in the world' (*Rabindranath Tagore: Poet and Dramatist*, second ed. p. ix). In Bengal Tagore's popularity is still immense; but there is something cloying to modern palates in the noble abstractions that press

218

upon each other in the English versions of the poems, and perhaps the short stories collected in *Hungry Stones* are easier to appreciate. A Pakistani critic has compared Tagore's novel *Gora* (white man) to *Kim*, rather to the latter's disadvantage, because Tagore resolves the problem of identification faced by the man whose father is English and mother Hindu while Kipling evades it (Syed Sajjad Husain, *Kipling and India*, pp. 161–3). Tagore's solution is for the son to become an Indian, dedicated to the service of all his fellow-countrymen and free to criticise racialism and bigotry wherever he finds them. But, in English at any rate, it is a dull book.

Fa-hien and Hiuen Tsang are only the first of many who have written of their travels in India, and from the 17th century onwards there are many books in Portuguese, Dutch, French and English that are still enjoyable to non-specialists. The place-names and vocabulary may be unfamiliar however, and anyone setting out to read such books would do well to get a copy of Yule & Burnell's *Hobson-Jobson*, first published in 1886, with a new edition by William Crooke in 1903. This describes itself on the title-page as 'a Glossary of Anglo-Indian Colloquial Words and Phrases, and of kindred terms; etymological, historical, geographical, and discursive', and in the preface Colonel Yule explains the title as a 'typical and delightful example of that class of Anglo-Indian argot which consists of Oriental words highly assimilated, perhaps by vulgar lips, to the English vernacular', and peculiarly apt in so far as it conveys a hint of dual authorship. The words are 'an Anglo-Indian version of the wailings of the Mohammedans as they beat their breasts in the processions of the *Moharram* – "Ya Hassan! Ya Hosain!"'

As in the O.E.D., definitions are supported by quotations, sometimes quite long ones such as the story of the post-runner and My Lord the Tiger from T.H. Lewin's *A Fly on the Wheel, or How I helped to Govern India* (1885), an excellent book. In his *Despatches* (1804), when Wellington refers to 'the Moorish language' he means Hindustani, the common language of the country, because to the Portuguese all Muslims came from Mauretania and therefore those they found in India were called Mouros too. The Gentoos (gentiles) as a name for the Hindus is also a Portuguese legacy, though it is sometimes restricted to the Telugu-speaking subjects of the Vijayanagar kingdom. Bishop Heber, when in 1826 he wrote 'Though it was still raining, I walked up the Bohr Ghat, four miles and a half, to Candaulah', meant he walked up a pass, in what by extension has also come to mean a range of mountains. In the seventeenth century a factory meant a trading establishment at a foreign port or mart, and *Hobson-Jobson* lists 24 factories in western India, 16 on the eastern and Coromandel coast, and 13 (Chittagong being followed by a question mark) on the 'Bengal side' – which includes Agra and Lahore as well as Dacca!

One need not always agree with Yule & Burnell, any more than one always agrees with the brothers Fowler, but it is a pleasure to come across pithy digressions such as the following on the avocado or alligator pear:

> The praise which Grainger...'liberally bestows' on this fruit, is, if we might judge from the specimens occasionally met with in India, absurd. With liberal pepper and salt there may be a remote suggestion of marrow: but that is all. Indeed it is hardly a fruit in the ordinary sense. Its common sea name of 'midshipman's butter' is suggestive of its merits, or demerits....Its introduction into the Eastern world is comparatively recent; not older than the middle of last century. Had it been worth eating it would have come long before.

Though Kipling said Gunga Din would be found

> ...squattin' on the coals
> Givin' drink to poor damned souls

he did not need *Hobson-Jobson* to remind him that *bheesty*, the name of the man who pours water from a goatskin slung on his back, is derived from the Persian *bihishti*, a person of *bihisht* or paradise; and Yule & Burnell confirm his reputation, saying:

> No class of men (as all Anglo-Indians will agree) is so diligent, so faithful, so unobtrusive, and uncomplaining as that of the *bihishtis*. And often in battle they have shown their courage and fidelity in supplying water to the wounded in face of much personal danger.

In short, *Hobson-Jobson* provides 900 pages of succulent browsing, leading to such pastures as the Abbé Dubois's *Hindu Manners and Customs* (1817), Colonel Meadows Taylor's *Confessions of a Thug* (1839), William Sleeman's *Rambles and Recollections of an Indian Official* (1844) and Emily Eden's *Up The Country* (1866).

Everyone has read *A Passage To India*. E.M. Forster was himself dissatisfied with it but it is the book on which his reputation mainly rests. In *The Hill of Devi* (1953) he says:

'I began this novel before my 1921 visit and took out the opening chapters with me, with the intention of continuing them. But as soon as they were confronted with the country they purported to describe, they seemed to wilt and go dead and I could do nothing with them. I used to look at them of an evening in my room at Dewas and felt only distaste and despair. The gap between India remembered and India experienced was too wide. When I got back to England the gap narrowed, and I was able to resume. But I still thought the book bad, and probably should not have completed it without the encouragement of Leonard Woolf.'

In India Muslims, Hindus and the British all consider they have been maligned in it; but Indians find no difficulty in recognising the Burtons and Turtons, and Westerners have come across Azizes and Godboles. Forster's characters have the truth of caricatures, with the salient defects of the three communities exaggerated. This is the only novel about India that comes near to grappling with its complexity, and both for its descriptions – sunrise or the Gokul Ashtami celebrations – and Forster's quiddity, it will be read again and again. The following passage perhaps epitomises the author's bafflement:

India has few important towns. India is the country, fields, fields, then hills, jungle, hills, and more fields. The branch line stops, the road is only practicable for cars to a point, the bullock-carts lumber down the side tracks, paths fray out into the cultivation, and disappear near a slash of red paint. How can the mind take hold of such a country? Generations of invaders have tried, but they remain in exile.

221

The important towns they build are only retreats, their quarrels the malaise of men who cannot find their way home. India knows of their trouble. She knows of the whole world's trouble, to its uttermost depth. She calls 'Come' through her hundred mouths, through objects ridiculous and august. But come to what? She has never defined. She is not a promise, only an appeal.

Astronomy, physics, biology, chemistry and mathematics are presumably nationally neuter, but on anthropology, jurisprudence, economics, politics, sociology and current history there is a wealth of material which, even if I were competent to do so, there is no space to review. During the last twenty years much has been contributed by American scholars. Examples – with a good bibliography in Austin – are Selig Harrison's *India – the most dangerous decades* (1960); Granville Austin's *The Indian Constitution – cornerstone of a nation* (1966); and *The Crisis of Indian Planning* (1968), by Paul Streeten and Michael Lipton. A few useful titles on other subjects may be mentioned.

Michael Edwardes's *A History of India* (1961, enlarged 1967) is a good short introduction, with lively excerpts from contemporary sources and a Select Bibliography that covers five pages. He commends A.L. Basham's *The Wonder that was India* (1954) and Stuart Piggott's *Prehistoric India* (1950), which should now be supplemented by Bridget & Raymond Allchin's *The Birth of Indian Civilization* (1968), for excavation and carbon-dating have filled many gaps during the last twenty years. Another general history that may be singled out is *The Rise and Fulfilment of British Rule in India* (1934), by Edward Thompson & G.T. Garratt, and, for reference, *The Oxford History of India*, entirely the work of that giant, Vincent Smith, when first published in 1919, and reprinted in 1958 with the section on Ancient India revised by Sir Mortimer Wheeler and A.L. Basham, the Muslim period revised by J.B. Harrison and the British period rewritten by T.G.P. Spear.

To see the wild animals of India one has to visit sanctuaries such as Kaziranga or Kanha, but there are good pictures and descriptions in E.P. Gee's *The Wild Life of India* (1964), B.

Seshadri's *The Twilight of India's Wild Life* (1960), and S.H. Prater's *The Book of Indian Animals* (third edition, 1971). A big variety of birds however, many of them unfamiliar and brightly coloured, can be seen even in busy towns and visitors should carry the *Common Album of Indian Birds*, by Selim Ali and Laeeq Futehally, issued by the National Book Trust. More comprehensive books on birds are easily available too, such as Hugh Whistler's *Popular Handbook of Indian Birds* (fourth edition, 1949), and the two complementary books by Selim Ali, *The Book of Indian Birds* (eighth edition, 1968) and *Indian Hill Birds* (1949), or G.M. Henry's *Guide to the Birds of Ceylon* (second edition, 1971) which is quite serviceable for use in southern India. On trees, flowers, reptiles and insects there are no satisfactory popular books, though *Some Beautiful Indian Trees*, by E. Blatter & W.S. Millard (second edition revised by W.T. Stearn, 1954) and *Flowering Trees and Shrubs in India* by D.V. Cowen (fifth edition, 1969) contain some good illustrations. EHA's *The Tribes on my Frontier, an Indian naturalist's foreign policy* (1883) can still be read with amusement and profit. Jim Corbett's *Man-eaters of Kumaon* (1940) is widely known and his four other books are almost equally good.

To religion and philosophy S. Radhakrishnan's *The Hindu View of Life* (1927) is a good introduction and Heinrich Zimmer's *Philosophies of India*, edited by Joseph Campbell (1951), deals also with Buddhism and Jainism. The *Bhagavad Gita* (part of the epic *Mahabharata*) is, according to Zimmer, 'the most popular, widely memorized authoritative statement of the basic guiding principles of Indian religious life'; it is available in many translations, of which the easiest to understand is that by Christopher Isherwood and Swami Prabhavananda. An even more popular book in Hindi is Tulsidas's *Ramayana*, in F.S. Growse's translation of which Lockwood Kipling wrote:

a book which ought to be studied by all those who care to know what the best current Hindu poetry is like, in its sugared fancy, its elaborate metaphors, its real feeling for the heroic and noble, and also its tedious, unaccidented meandering.

223

On Islam in India – as indeed on many other subjects – *Modern India and the West*, edited by L.S.S. O'Malley (1941), can safely be recommended.

The opulent sketch-books of Indian scenery, monuments, birds, costumes and customs, such as those executed by Thomas, William and Samuel Daniell at the beginning of the nineteenth century, are not often seen except piecemeal, with their leaves cut out and framed. James Fergusson's *History of Indian and Eastern Architecture* (1876) and the *Cave Temples of India* written four years later in collaboration with James Burgess, have been reprinted however. A good selection of plates will be found in Vincent Smith's *History of Fine Art in India and Ceylon* (second edition 1935, reissued 1962 with slight revision by Karl Khandalavala) and Heinrich Zimmer's *Art of Indian Asia* (2 vols., second edition, 1960). There are many other well illustrated books on painting, sculpture and architecture. *Edward Lear's Indian Journal* (1953) contains extracts from his diaries and an all-too-niggardly selection from the 1500 sketches he made during his visit in 1873–4, and one has to hope that one day Harvard will publish a more substantial selection from their rich hoard. Desmond Doig's *Calcutta – An Artist's Impression* (1969) skilfully captures its former beauties and present squalor.

Autobiography has been surprisingly little pursued, in India but has produced three outstanding examples.* M.K. Gandhi's *The Story of My Experiments with Truth* (1927–9) covers only the early part of his life and even then needs to be supplemented by *Hind Swaraj* (1909) and *Satyagraha in South Africa* (1928). Jawaharlal Nehru's *Autobiography* (1936), an equally candid though less introspective book, also covers only the early period. Nirad Chaudhuri's *Autobiography of an Unknown Indian* (1951) describes his boyhood and youth in Bengal. All three are fascinating. John Masters's *Bugles and a Tiger* (1956) gives a vivid picture of army life before the second world war and *The Tribal World of Verrier Elwin* (1964) covers the whole life of this philanthropologist – the label is his own – who devoted himself to recording the fast-vanishing ways of

---

*A partly autobiographical book, subtitled 'An Exploration of India', is *An Area of Darkness* by V.S. Naipaul whose family moved from Uttar Pradesh to Trinidad two generations ago, which gives an illuminating account of modern India. Eds.

life of the aboriginal peoples of central and north-eastern India.

What makes people laugh varies greatly from area to area and time to time, and the misuse of English idiom – 'in this way my whole career has been blighted; and it is Mr Justice X. who is the blighter' (quoted from SPE Tract XLI, 'Some Notes on Indian English' by R.C. Goffin) – has added the word babuism to the English language. Such babuisms are often apt as well as funny, like Poirot's idiotisms, and they are the mainstay of much that is meant to be humorous. R.K. Narayan is often very funny and I confidently recommend *The Financial Expert* (1952).

The first detective in English fiction, Sergeant Cuff, appears in Wilkie Collins's *The Moonstone* (which was in fact a diamond from Seringapatam) but he carries out his investigations in England and there seem to be very few detective stories with an Indian setting. A.E.W. Mason's *The Broken Road* (1907) is a straight novel about the north-west frontier. 'Philip Woodruff' (Philip Mason, chronicler of the Indian Civil Service in *The Founders*, 1952, and *The Guardians*, 1954), wrote well of police methods in an Indian village in *Call the Next Witness* (1945), and recently H.R.F. Keating has created Inspector Ghote of the Bombay C.I.D.

The most prolific contemporary Indian novelist is probably R.K. Narayan, who has created Malgudi, a small town not far from Bangalore, and made us friendly towards its varied and amusing population. Raja Rao's *Kanthapura* (1938), written in a highly individual style attempting to reproduce his native Kannada, shows the impact of Gandhi's non-co-operation movement on a south Indian village. Mulk Raj Anand's *Untouchable* (1935), Ahmed Ali's *Twilight in Delhi* (1940), John Masters's *Nightrunners of Bengal* (1951), Khushwant Singh's *Train to Pakistan* (1956), Manohar Malgonkar's *Distant Drum* (1960), Attia Hosain's *Sunlight on a Broken Column* (1961), Ruth Prawer Jhabvala's *The Householder* (1963) and Paul Scott's *The Jewel in the Crown* (1966) stand out among recent books of fiction, and comprehensive lists of novels in English about India are to be found in Bhupal Singh's *Survey of Anglo-Indian Fiction* (up to 1930), K.R. Srinivasa Iyengar's *Indian Writing in English* (up to 1960),

225

M.E. Derrett's *The Modern Indian Novel in English* (1947–64) and Allen J. Greenberger's *The British Image of India: A study in the literature of imperialism* (1860–1960).

Appendix

Notes on Contributors

JOHN KENNETH GALBRAITH Paul M. Warburg Professor of Economics at Harvard University, he was President Kennedy's Ambassador to India from 1961–63, during which time Jawaharlal Nehru said of him: 'he is an able and brilliant man and we are all grateful for the help he has given India during these last years'. Among his many books are *The Affluent Society*; the *New Industrial State*; *Ambassador's Journal: A personal account of the Kennedy years*; *Economics, Peace and Laughter* and *Indian Painting: The Scene, Themes and Legends*. His wife Catherine is herself the author, with Rama Mehta, of a history of India, *India: Now and Through Time*. Professor Galbraith is currently President of the American Economic Association.

EDWARD HOWE. A British journalist who lived and worked in India for fourteen years as head of a British news agency, is married to Robin Howe. His work has carried him to almost every corner of the subcontinent, mostly by car. He has motored to India twice from Britain.

KHUSHWANT SINGH. Studied law at London University and turned to writing as a profession after a period of teaching Hindu Law at the University of Punjab, and serving in a number of diplomatic posts in Canada, London and New York. He is now editor of one of India's largest circulation illustrated weeklies. He is the author of several novels but best known for his history of the Sikhs.

FRANK MORAES. One of India's senior editors, Frank Moraes

227

has been well placed to watch history unfolding in India both
before and after Independence. His biography of Jawaharlal
Nehru is a standard work. The keen and critical approach
which he gives to the present Indian political scene is respected
by both politician and observer.

JAMES CAMERON. Born in 1911, he was educated erratically at
various schools, mostly in France. He has worked as a
journalist since 1928 and travelled to most parts of the world
as a foreign correspondent. He appears frequently on television, for which medium he has produced numerous documentaries on contemporary subjects. He won the Granada Award
for the Journalist of the Year in 1965, was chosen Granada
Foreign Correspondent of the Decade in 1966 and, in the same
year received the Hannen Swaffer Award for journalism. He is
the author of several books, including *Touch of the Sun,
Witness in Viet Nam* and *Point of Departure.* He is married to
Moneesha Sarkar, who works for Air-India, and lives in
London.

PAUL SCOTT. His first novel, *Johnnie Sahib,* won the Eyre &
Spottiswood Literary Award in 1951. This was followed by
*The Alien Sky, A Male Child, The Mark Of The Warrior, The
Chinese Love Pavilion* and *The Bird Of Paradise,* all published
by Eyre & Spottiswoode and reissued by Heinemann. Then
came *The Bender* and *The Corrida At San Feliu* published by
Secker. In 1966 Heinemann published his great bestseller, *The
Jewel In The Crown* and in 1968 its sequel, *The Day Of The
Scorpion.* His latest book, published in 1971, is the third in the
Raj quartet and it is entitled *The Towers Of Silence.*

The late MANOHAR MALGONKAR. Born in 1913, he spent his
boyhood in Jagalbet, a village deep in the Kanara jungles of
Western India. After graduating from Bombay University, he
became a professional big-game hunter. Within two years,
however, he gave up 'killing wild animals for a living', went
into government service, and never killed a wild animal again.
He was a 'furious' conservationist. During the war, his department 'loaned' him to the Army and he reached the rank of
Lieutenant-Colonel. He began writing to supplement his army

pay, and on leaving the army in 1952, became a regular contributor to Indian radio and magazines. He lived in his jungle village, dividing a ten hour working day between writing and business. Author of two historical works, *Kan Hoji Angrey* and *The Puars of Dewas Senior*, he also wrote a novel about the Indian Army, *Distant Drum*. He left a widow and one daughter.

KAMALA MARKANDAYA. Born and brought up in South India, she attended a number of schools there, and later Madras University. Her interest in villages stems partly from her background (her family were landowners) and partly from an experiment in rural living which she undertook out of curiosity to discover how the other half lived. Later she worked as a journalist. Her first book was based on her rural experiences. Other novels include *A Handful Of Rice, The Coffer Dams* and, most recently, *The Nowhere Man*. She now lives and works in London, and her return trips to India always include visits to village communities which remain an abiding interest. She is married and has a daughter.

C.C. JOSEPH. Has spent many years researching into Indian religions and culture, as well as astrology and palmistry. Hailing from Kerala, he now lives in New Delhi and is a leading member of the capital's Malayalam community. Besides speaking Malayalam, in which language many of his poems have been published, he also has a knowledge of Sanskrit, Tamil and Hindi, but has practised as a journalist writing in English for most of his life. Mr Joseph is married and has three children and several grandchildren. He also has a vine in his New Delhi garden which repays his devoted attention by producing grapes most years.

MULK RAJ ANAND. As Chairman of the Lalit Kala Akademi (National Academy of Art) in New Delhi, and editor of *Marg*, an art magazine published in Bombay, Mulk Raj Anand is an acknowledged judge and critic of India's art heritage. His life has been spent in studying the carvings of Indian temples and the various ancient schools of painting. He is also a novelist of note.

MAXWELL FRY C.B.E. Professor of Architecture at the Royal Academy and a Fellow of the Royal Institute of British Architects. He won the Royal Gold Medal for Architecture in 1964. In a long and distinguished career he has practised with Walter Gropius, Le Corbusier and Jane Drew, who is also his wife. He was Senior Architect responsible for Chandigarh, the new capital of the Punjab, and has built many buildings in England, Ghana and Nigeria. He is the author of *Fine Building* (jointly with Jane Drew), *Architecture for Children* and *Art in a Machine Age.*

TARA ALI BAIG. The wife of a former Indian diplomat, she was educated in Darjeeling, Switzerland, the United States and Dacca. As a social worker in India, she came close to the people and the problems of the subcontinent, and her interest in the traditional background of the country led to the publication of her book *Women in India,* a source book on Indian women. She has also written a novel, *Moon in Rahu,* based on the famous Bhowal Sannyassi case, and is the author of the official biography of Sarojini Naidu, politician, poetess and friend of Gandhi. Feeling that the costumes of India are an essential part of India's history not much dealt with by historians, she has applied her professional skill to a study of the subject which helps fill a gap. Mrs Tara Ali Baig has contributed many articles to leading Indian journals, and has written for *The Times* of London and *UNICEF Carnet d'Enfance.*

INDRANI. One of the most distinguished names in classical dance in India. The daughter of the famous dancer Ragini Devi, she has been dancing since childhood. She performs Bharata Natyam and Mohini Attam, and has been responsible for the revival of Kuchipudi dance. She also pioneered the revival of *Orissi,* which she has taken out of Orissa and round the world. For this she was awarded the Padma Shri, for her contribution to the arts, by the Indian Government in 1969. She has danced in all five continents of the world and lauched numerous young dancers on their careers in India.

AMITA MALIK. Film critic of *The Statesman,* New Delhi, and Indian correspondent of *Sight and Sound,* London, is widely recognised as India's most distinguished film correspondent. She travels widely and is a familiar figure at international film festivals. She has interviewed most directors, from Ingmar Bergman to Kurosawa, and her deep knowledge of both Indian and international films makes her an ideal interpreter of the Indian film scene for both Indians and Westerners. She has appeared in this role on many TV networks throughout the world and has also written for *The Times* of London and the *Frankfurter Allgemeine Zeitung*.

ROBIN HOWE. A writer on international cookery who has travelled widely throughout the world in search of good and interesting food. One of the latest of her eighteen books on international cuisines concerns the history of food in India, a subject which interested her during a long period of living in India.

The late E.P. GEE. An Englishman who spent more than half his life in India, much of which was dedicated to recording wild life, both in pictures and words. A new species of golden langur was named after him by a grateful Indian Government. He travelled extensively in search of bird, beast and reptile and did much to interest others in the world of wild life and, we hope, save much of it for future generations.

PETER JACKSON. Became interested in birds while on a two-month trek in the Himalayas to report the successful British Everest expedition in 1953 for Reuter's news agency, for whom he was Chief Correspondent in Delhi for 18 years. Watching and photographing birds – and animals – became his off-duty passion, and his work gave him opportunities to pursue his hobby in many parts of southern Asia. An active member of the Delhi Bird Watching Society, of which he became Honorary Secretary, he added a number of species to the area bird list and contributed notes to the Journal of the Bombay Natural History Society. Interest in birds led to his involvement with conservation and the decision to take on the post of Director of Information of the World Wildlife Fund to advance the cause.

R.E. HAWKINS. Went to India as a young man and has spent more than forty years working there in publishing, the latter part as head of the Oxford University Press. He was responsible for discovering many writers and encouraging them to describe India and the many facets of life there. Although now retired, he still lives in India.

A Note on India Tourist Offices

There is a good network of Indian Government Tourist offices both in India and abroad. These are non-commercial organisations and cannot make travel arrangements, but the staff have all the information you will need for your journey.

A Note on Wild Life Sanctuaries

Below is a list of animal and bird sanctuaries which may be visited by tourists to India. There is a great deal of information about guides, transport, accommodation, closed seasons, etc. which visitors should obtain from Indian Tourist Offices abroad or on the spot.

Gir Forest, Gujarat State.
The Taroba National Park, Maharashtra State
Keoladeo Ghana Sanctuary, Rajasthan
Sariska Sanctuary, Rajasthan
Jai Samand Sanctuary, Rajasthan
Dachigam Sanctuary, Kashmir
Corbett National Park, Himalayan foothills
Chandraprabha Sanctuary, Uttar Pradesh
Kanha National Park, Madhya Pradesh
Shivpuri National Park, Madhya Pradesh
Hazaribagh National Park, Bihar
Jaldapara Sanctuary, West Bengal
Kaziranga Sanctuary, Assam
Manas Sanctuary & North Kamrup Sanctuary, Assam
Bandipur Sanctuary, Mysore
Ranganthittoo Bird Sanctuary, Mysore
Mudumalai Sanctuary, Tamilnad
Vedanthangal Bird Sanctuary, Tamilnad
Periyar Sanctuary, Kerala

EUROPE

SUEDE
MER BALTIQUE
LAPONIE
MER
NOUVELLE
ALLEMAGNE
PRUSSE
LIVONIE
MOSCOVITE
HONGRIE
POLOGNE
RUSSIE
GRANDE TARTARIE
SIBERIE

MER
MEDITERRANEE
GRECE
TURQUIE EUROPEANE
MER NOIRE
CIRCASSIE
CALMOUCS NOIRS
MAQUIE PAYS DES CALMO
GRANDE TARTARIE
INDE

TURQUIE ASIATIQUE
NATOLIE
GEORGIE
CALMOUCS BLANCS
ROYAUME DE
ou DE L'AYOUC
Horde de Kontaicin
Horde de Kassabin
Horde de Koltuei
TARTARIE
TURQUE
ROYAUME DE
DE COTAN

EGYPTE
SYRIE
Desert
IRAC
PERSE
CHORASAN
PAYS DES USBECS
ROYAUME DE LAB
ROYAUME DU CO
ROYAUME DE CO

ARABIE DESERTE
ETAT DU CHERIF
IRAC ARABI
FARSISTAN
KERMAN
SABLESTAN
MECRAN
Chaparanque

MER
ROUGE
ARABIE
BAHREIN
OMAN
Mascate
ETATS
DU MOGOL
ROYAUME DE BE
ROY. D'ORIXA

NUBIE
YEMEN ou ARABIE HEUREUSE
Hadramaut
Aden
Isle de Socotora
Cap Guardafui

ABISSINIE
ROYAUME DES GALLES
ROY. D'ADEL
AJAN
Coste d'Ajan
Coste des Negres

LAQUEDIVES
Cap Comorin
ISLE DE CEILAN

Equateur ou Ligne Equinoctrale

MALDIVES independantes
Isle de Diego Ruys
Diego Rodriguez

MER DES

CARTE D'ASIE
dressée sur les Memoires envoyez par le Czar
a l'Academie Royale des Sciences
Sur ce que les tenebres sont été laissés de plus exact des pays orientaux
Sur un grand nombre de Relations de terre et de Mer et de Cartes
manuscrites detaillées. Le tout augmenté aux Observations de l'Academie
establies de R.R.P.P. Jesuits et autres Mathematiciens. Par G. Del Isle.
Nouvellement corrigé après les derniers Decouvertes
faite par l'Academie de Petersbourg.
A AMSTERDAM
Par COVENS & MORTIER